Understanding Interfaces

Computers and People Series

Edited by

B. R. GAINES and A. MONK

Monographs

Communicating with Microcomputers. An introduction to the technology of man–computer communication, *Ian H. Witten* 1980
The Computer in Experimental Psychology, *R. Bird* 1981
Principles of Computer Speech, *I. H. Witten* 1982
Cognitive Psychology of Planning, *J-M. Hoc* 1988
Formal Methods for Interactive Systems, *A. Dix* 1991
Human Reliability Analysis: Context and Control, *E. Hollnagel* 1993

Edited Works

Computing Skills and the User Interface, *M. J. Coombs and J. L. Alty (eds)* 1981
Fuzzy Reasoning and Its Applications, *E. H. Mamdani and B. R. Gaines (eds)* 1981
Intelligent Tutoring Systems, *D. Sleeman and J. S. Brown (eds)* 1982 (1986 paperback)
Designing for Human–Computer Communication, *M. E. Sime and M. J. Coombs (eds)* 1983
The Psychology of Computer Use, *T. R. G. Green, S. J. Payne and G. C. van der Veer (eds)* 1983
Fundamentals of Human–Computer Interaction, *A. Monk (ed)* 1984, 1985
Working with Computers: Theory versus Outcome, *G. C. van der Veer, T. R. G. Green, J-M. Hoc and D. Murray (eds)* 1988
Cognitive Engineering in Complex Dynamic Worlds, *E. Hollnagel, G. Mancini and D. D. Woods (eds)* 1988
Computers and Conversation, *P. Luff, N. Gilbert and D. Frohlich (eds)* 1990
Adaptive User Interfaces, *D. Browne, P. Totterdell and M. Norman (eds)* 1990
Human–Computer Interaction and Complex Systems, *G. R. S. Weir and J. L. Alty (eds)* 1991
Computer-supported Cooperative Work and Groupware, *Saul Greenberg (ed)* 1991
The Separable User Interface, *E. A. Edmonds (ed)* 1992
Requirements Engineering: Social and Technical Issues, *M. Jirotka and J. A. Goguen (eds)* 1994

Practical Texts

Effective Color Displays: Theory and Practice, *D. Travis* 1991
Understanding Interfaces: A Handbook of Human–Computer Dialogue, *M. W. Lansdale and T. R. Ormerod* 1994

EACE Publications
(Consulting Editors: *Y. WAERN and J-M. HOC*)

Cognitive Ergonomics, *P. Falzon (ed)* 1990
Psychology of Programming, *J-M. Hoc, T. R. G. Green, R. Samurcay and D. Gilmore (eds)* 1990

Understanding Interfaces

A Handbook of Human–Computer Dialogue

Mark W. Lansdale

and

Thomas C. Ormerod

Department of Human Sciences
Loughborough University of Technology,
UK

ACADEMIC PRESS

Harcourt Brace & Company, Publishers

London   San Diego   New York
Boston   Sydney   Tokyo   Toronto

ACADEMIC PRESS LIMITED
24–28 Oval Road
LONDON NW1 7DX

U.S. Edition published by
ACADEMIC PRESS INC.
San Diego, CA 92101

This book is printed on acid-free paper

A catalogue record for this book is available from the British Library

ISBN 0-12-528390-3

Typeset by Phoenix Photosetting, Chatham, Kent
Printed in Great Britain by Hartnolls Ltd., Bodmin, Cornwall

Contents

Preface

The title 'Understanding Interfaces' has a dual interpretation. First, there is a wealth of research and development work associated with the design and use of machine interfaces which requires a detailed explanation. Thus, one purpose of this book is to outline the nature and design of interfaces based around the available computing technologies. Second, as researchers into the use and design of interfaces, we are interested in developing interfaces which are constructed in a way that is sympathetic to their potential users. Thus, another purpose of this book is to discuss the extent to which designers can develop interfaces that 'understand' their potential users.

There are a number of potential structures that might be used to organize a textbook on interface use and design. One might choose to have a chapter on each of the components of an interface (input/output devices, display issues, dialogue styles, etc.). Alternatively, one might choose to have a chapter on each domain of interface use (process control, computer programming, vehicle displays, etc.). One might even structure the book around the psychological characteristics of users (perception, attention, memory, problem solving, etc.). Each of these structures has advantages and disadvantages, but tends to emphasize one viewpoint, which may be a technological, managerial or psychological analysis, at the expense of others. In order to try and capture and integrate the contributions to understanding interfaces that each structure offers, we have divided the book into four parts. These correspond roughly to the following topics: an introduction to the issues of interface use and design; understanding interfaces in terms of human communication; understanding the skills necessary for interface use; and understanding the design and evaluation of interfaces.

Part One   Interfaces in context

This part contains only one chapter which sets out the context for understanding interfaces and serves to lay out a number of issues which run through the book. The relationship between humans and the devices they develop is described, and some of the economic, social and political factors which influence the use and development of interfaces are introduced. A brief overview of cognitive processes is given, and a guide is given as to the ways in which different aspects of human performance are discussed throughout the book.

Part Two   Understanding dialogue

In the four chapters of Part 2 we discuss different aspects of dialogue for human-computer interaction. Here the term 'dialogue' is extended to include the idea that the flow of information between the user and the interface is equivalent in many respects to human-human dialogues. The idea that human-human dialogues are a good model for the design of human-machine interfaces is a controversial topic, with arguments being made strongly both for and against the idea that interfaces should be designed to emulate natural language constructions (e.g., Fitter, 1979). The problem is to decide which aspects of human-*human* dialogue are sensible candidates for emulation in human-*computer* dialogue. In this part of the book we consider the *medium* through which a dialogue is conducted and the *task* it is intended to achieve. If these are understood, as well as the psychological and sociological aspects of human-computer interaction, then a clearer definition of what dialogue is emerges, and with it a better understanding of the extent to which human-computer and human-human dialogues should be seen as comparable.

In Chapter 2, *Fitting dialogues to the task*, we discuss how different styles of human-computer dialogue are optimized for different tasks, and how their adequacy also depends on the level of expertise of the user.

Chapter 3, *Fitting dialogues to the medium*, outlines the technologies of human-computer communication and the ways in which different media might facilitate or constrain channels of communication.

In Chapter 4, *The structure of dialogues*, we look at the linguistic elements that make up human-computer dialogues. In particular, we focus upon the ways in which artificial command languages are constructed.

Chapter 5, on *Maintaining dialogues*, is concerned with the methods whereby users, or by good design the interface itself, can control a dialogue. This chapter is last in this sequence because it represents the coming together of the elements of dialogue described in the previous chapters.

Part 3   Understanding user skills

This part of the book considers what we mean by 'user skills', and we examine how they might be acquired or modified to meet new circumstances and tackle new interfaces. There is a greater emphasis on psychological theories than in previous chapters simply because the focus is more on users and less on tasks and technologies. Each chapter presents a different view of user skill.

In Chapter 6, *Skill as procedures*, we describe how some user skills are based on learned procedures for interacting with interfaces, and we examine the way in which users can acquire such skills through extensive practice.

In Chapter 7, *Skill as understanding*, we examine the role of prior knowledge and expectations in shaping users' understanding of interfaces, and we discuss the validity of designing interfaces which are consistent with users' experience or conceptual understanding.

In Chapter 8, *Skill as exploration*, we describe how users can acquire knowledge through exploring an interface, and a number of current vogues for enabling the exploration of interfaces are discussed (such as object-oriented systems, direct manipulation and hypermedia).

Part Four   Understanding interface design

Whilst Parts 2 and 3 of the book are essentially concerned with the psychology of the user, in Part 4 we consider the practical issues surrounding the design and evaluation of interfaces. The purpose of this section is not to provide a detailed practitioner's guide to interface design, but to enable readers to understand the context in which methods for design and evaluation might best be applied. To this end we describe a number of formal and informal methods used in the design and evaluation of interfaces, examining the reasons why these methods might be applied, and comparing their relative advantages and disadvantages.

In Chapter 9, *Designing interfaces*, we describe the processes involved in the design of interfaces, and examine the means by which specialists can influence this process. A central theme in this chapter is the role of task analysis as an essential part of interface design.

In Chapter 10, *Evaluating interfaces*, we conclude the book by discussing the evaluation of interfaces, and we assess the concept of 'usability'. We describe how different methods of evaluation may be compared on the basis of a cost-benefit analysis. We argue that to choose an appropriate evaluation method requires an understanding of the context in which an interface will be designed and used. To develop a sufficient understanding of this context requires detailed knowledge of the tasks, technologies and users that are outlined in the rest of the book. Thus the final chapter serves to draw together the different perspectives offered throughout the book, all of which are required to understand interfaces.

Acknowledgements

This book has been written over the past two years, and we owe an enormous debt of thanks to the many people who have supported us in this task. Our colleagues, in particular Andrew Shepherd and Linden Ball, have discussed many of the chapters in depth and have helped to shape the structure of the book significantly, and thanks are due to Marion Gibson and Rachel Fairhurst for their assistance in typing and organizing the text. The text was read and commented upon in draft form by a number of people, including Adrian Bailey, Thomas Green, John McEwan and Richard Whitehand. We are indebted to Andrew Carrick, editor at Academic Press, for his guidance, encouragement and patience. We were so bad at meeting deadlines that Andrew had moved on from Academic Press before the book was completed. However, we were lucky enough to be supported through the latter stages of writing by Kate Brewin, who finally got us to come across with the copy. We thank also the series editor Andrew Monk and the three anonymous referees who provided valuable feedback throughout and helped us to improve the book in many ways.

"*Interface* is a metaphor that has become increasingly common in official writing. If one or two dimensions are enough for you, you can make do with a *point of contact* or a *common frontier*. If you hanker after a third dimension, you will feel the need of an *interface*. It is at present a vogue word and must be regarded as on probation. The fact that it is sometimes 'broad', sometimes 'virtual' and sometimes sat on ('I find myself sitting on a number of interfaces') suggests that it may need watching."

Definition from *The Complete Book of Plain Words*, 3rd Edn.,
Penguin Books, 1986, page 236.

Part 1
Interfaces in Context

In this part of the book we discuss the context in which interface design has become an important issue in the development of technology used by individuals and organizations. We define what is meant by the term 'interface', and discuss the basics of cognitive psychology (how our minds process information) and how they apply to the use of interfaces; we consider the criteria of safety, effectiveness and profit in the design of interfaces; and finally, we introduce the design of interfaces.

CHAPTER 1

The context of interface use

Overview

In this chapter we describe the main characteristics of interfaces, and introduce the topic of human-computer interaction. To understand why an interface does or does not work effectively, it is necessary to examine interface technologies, their users, the tasks to be supported by the interface, and the environment in which users will carry out these tasks. This chapter begins the process of explaining how these topics each contribute to the use of interfaces, a process which is continued in the rest of the book. We examine the role that psychology can play in the study and design of interfaces, and we also look at the broader context of interface design, and the influence that factors such as safety and cost have on the design of interfaces.

1.1   What is an interface?

The Shorter Oxford English Dictionary (OED) defines the word 'interface' as follows:

1. A surface separating two portions of matter or space and forming their common boundary.
2. A place or region, or a piece of equipment, where interaction occurs between two systems, organizations or persons.

It is perhaps significant that 'interface' scrapes into the OED only as part of the addenda. It is also interesting to note its increasing misuse as a verb, where to 'interface' with other people or computers is to converse with them. The word 'interface' clearly describes an emergent and rapidly-changing subject, and requires further definition to capture the sense in which we use it in this book.

The two definitions offered by the OED suggest important, but not entirely consistent, views of interfaces. The first suggests a boundary between humans and the machines they use. It is appropriate to use an analogy of a geographical boundary such as a river or mountain range to describe this interface, since it is frequently seen as a difficult boundary to cross. There is a range of limitations, both of people and products, which prevents people making the most effective use of everyday machines like ovens, washing machines, cash dispensers and electric typewriters. Seeing interfaces as a barrier is a common, if regrettable, view of technology. One message of this book is that it is avoidable.

The second definition, which includes 'a region ... where interaction occurs', is a more positive use of the word. It introduces the notion of reciprocal action between humans, other humans, and devices. Another central theme of this book is the potential that a well-designed human-computer interface offers for enhancing creative and effective work and leisure practices. The key concept here is *reciprocity*: maps, road signs and books might be considered to be types of interface in that they communicate information, but they are passive devices. Devices such as video recorders have uses and performance which go beyond what can be seen. They can store sequences and can be in a number of different 'modes'. Pressing a key in one condition produces different results to pressing the same key in other circumstances. In short, they are proactive rather than passive devices, and it is this characteristic which distinguishes simple interfaces (such as maps) from the devices which are the focus of this book.

We refer to an interface as comprising the collection of objects, tools, languages and displays which lies between people and the machines that they intend to use. This definition might apply to a vast range of things, such as computer and video screens, telephone switchboards, word processing keyboards, factory control panels, car dashboards, computer programming languages, and so on. An interface is not just a physical entity. It can include symbols, concepts or words. For example, the command set of a word processor (e.g., selecting the Control and D keys to delete a block of text) is an interface, since it is the reciprocal understanding of these commands by human and computer which enables the task of editing a document to be carried out. Whilst there is great variation in the nature of things that might comprise an interface, they all have one property in common: they are media through which information is communicated. More specifically, they are used to communicate instructions, or to receive the results of implementing instructions.

Consider, as an example, the domestic videotape recorder. Most modern video recorders supply a hand-held remote control unit which can be used to operate play-back of pre-recorded video tapes or to instruct the video system to record a programme in advance. The buttons on the remote control are part of the interface used to issue a set of instructions (choosing the channel, setting the date, time and duration to record, etc.). The symbols drawn on the remote control are an important part of the interface, since they convey the purpose of each button. There is also (usually) an LED display on the video recorder which registers the instructions issuing from the remote control, thereby communicating the current state of the video recorder to its owner. On sophisticated video systems there may also be an on-screen display to guide the programming of the video recorder. So why is it that, given all these interface components at your disposal, when you want to record an episode of a favourite drama series whilst at a friend's house, you come home to find a tape full of news broadcasts and sport?

There is no doubt that many, if not most, of the interfaces which surround us in our everyday lives have hidden complexities, some of which come to light only when we make an error. In this example, a subtle aspect of a video recorder system interface is the sequence in which instructions must be issued to the video recorder. The requirement to order instructions in a form that is comprehensible to the video recorder's circuits is perhaps the main source of difficulty encountered by people trying to pre-record programmes. To alleviate this problem, some manufacturers issue a remote control which contains a bar code reader (like those used in supermarkets). Using this device, one can set the video simply by dragging the reader across bar codes printed alongside the required programme in a TV listings magazine. This is a clever technological solution to the problems of interface complexity, in that the task of setting a video recorder no longer requires one to learn a sequence of instructions. However, it has a serious drawback in that times of programmes are frequently changed to make room for unscheduled coverage of major news and sport events, thereby making the pre-defined bar codes inaccurate. Whilst it is usually possible to revert to a traditional method of programming the video recorder, this raises an additional problem that reliance on the bar code reader has removed the requirement to learn and practice the necessary sequences of instructions. Therefore, the hapless video owner must attempt an unfamiliar task at short notice and without much room for error. Our example illustrates a basic tension in the use of modern technology: either one must grapple with interfaces which can be complex, hard to use, and/or require training; or one must tolerate the loss of flexibility that often comes with simpler methods.

Specifically, this book is about understanding human-computer interfaces. Human-computer interaction is a growing field of study, and most research into the design and use of interfaces is in this domain. Why concentrate upon computers? There are two main reasons:

a) *Computers figure in many products.* As recently as 20 years ago computers were specialized devices often sitting in a room of their own doing a pre-determined job. Their role might have been comparable to an oscilloscope on an electronics workbench. Today they are a necessary part of most systems, including medical monitoring equipment, cars, and even washing machines. They are still used by computer scientists for programming and numerical analysis, but they are also used by pilots, deep-sea divers, and waiters in restaurants. The understanding of human-computer interfaces has become a central issue in the design of a wide range of products being used by more and more people.

b) *Human beings are an unchanging element in human-machine systems.* The development of computers has not suddenly revealed a range of previously unforeseen psychological phenomena specific to using computers. The issues involved in the design of human-computer interfaces apply equally to any other interface, be it a road sign, cockpit display, or knitting pattern. This is not to say that the information processing that occurs when using interfaces is necessarily the same for different tasks, but that the *constraints* that the human cognitive system imposes are the same. While many issues have been thrown into new focus by the spread of computing into different areas of work and leisure, most of them are not new. One can look back at research in the areas of process control (e.g., Crossman, 1974) or aeroplane flight-deck design (e.g., McFarland, 1953) to see the same issues being studied today, albeit in a different context and with a more sophisticated theoretical framework. To study human-computer interaction encompasses most of the issues in applied psychology.

Whilst this book discusses the physical objects which comprise interfaces, it is also about the aspect of interfaces we cannot see: the human mind. It is concerned with the psychological issues inherent in the devices we use, the screens we look at, and in the way our minds work. As much as anything else, it is these psychological issues which determine whether an interface will be seen as well-designed and efficient. Users of interfaces differ in many complex ways. Consider, for example, the design of video recorders. It is reasonable to assume that manufacturers generally deliver video recorders which are technically operable. The problems arise when the remote control leaves the showroom and is plunged into the hands of someone who didn't design, build or sell the video system, but will nonetheless expect to spend a minimal amount of time reading the accompanying manual before being able to record TV programmes. The users of devices like video recorders vary in many important respects that determine whether an interface is satisfactory for the tasks intended by the manufacturers. For example:

- Users vary in literacy, and will understand the accompanying instructions to a greater or lesser degree.
- Users vary in what they can remember about instructions.
- Users vary in how they interpret symbols drawn on the buttons (e.g., does < mean forward or backward?).
- Users vary in patience and enthusiasm to learn new ways of doing things.
- Users vary in prior experience of using technology.

The list could carry on almost indefinitely. The point is that the characteristics of a machine's users will determine the success of a machine's interface. These variations in user competence give rise to a number of issues concerned with training, dealing with users' errors and designing interfaces to match users' existing skills. These themes recur throughout the book. However, it would be misleading to concentrate purely upon the way people differ. In the next section we consider the commonalities of human performance. To introduce this, it helps to consider human beings in their evolutionary context.

1.2   Human performance and interface use

The evolutionary context of human performance

If you asked most adults how easy they found dessert spoons to use they might wonder why you asked. Such things are so well-designed for their purpose that we take them for granted. It is only when objects are hard to use or lead to errors or breakdowns in performance that we focus upon them. For example, Norman (1988), in his book *The Psychology of Everyday Things*, describes how even something as ordinary as a door to a building can present problems when it is not obvious at first glance how to open it. Some of these breakdowns are physical or physiological in origin: an information display might be obscured by glare (e.g., Cushman and Crist, 1987) or a particular keyboard configuration might cause some form of physical strain (e.g., Grandjean, 1987). However, many, if not most, breakdowns are psychological in origin. Despite the presence of technically adequate information and the means to communicate at the interface, users' inefficiencies can often be accounted for by some psychological limitation.

Discussions of human mental capacities are often rather negative in character because cognitive failures can have serious consequences for interface use. They concentrate upon topics such as our limited memory capacity, difficulties of learning, slips of action, and failures of attention. Undoubtedly all these limitations are genuine. Indeed, the purpose of this book is partly to discuss how they can be ameliorated in design. It is easy to make very unfavourable comparisons between humans and computers. Table 1.1 offers a comparison of human and computer performance at common tasks (encountered in a process control setting) that exemplifies such a view.

Despite the less than encouraging comparison offered in Table 1.1, most process plants do rely on human operators. This is partly because of the expense of automating plants so that they can be entirely controlled by computers. Mainly it is because there are some tasks which human operators can do (e.g., fault diagnosis) for which it is difficult if not impossible to develop computer-controlled mechanisms.

Table 1.1 Comparisons between human and computer performance

Facility	Human	Computer
Calculation (e.g., find the average daily flow rate from readings taken every 30 minute)	Slow, error-prone	Fast, accurate
Vigilance (e.g., monitor a level indicator to check for deviations in outflow volumes from a tank)	Intermittent, prone to fatigue	Permanent
Judgement (e.g., diagnose the cause of a pump failure)	Biased	Predictable

It is, therefore, misleading to see human performance in purely negative terms. Human beings have evolved successfully in a complex and hostile environment. As creatures lacking power, speed or massive reproductive turnover, we have relied for survival upon our ability to learn and adapt. But the world is so complex that no event is ever exactly the same twice. If you are trying to learn how to behave appropriately in the future, how do you know which aspects of future events to pay attention to and which to ignore? The answer is that humans solve this problem by selecting and interpreting only that information which, on prior experience, we assume to be relevant.

From an evolutionary point of view, the advantage of selecting and interpreting only relevant information is that it reduces the amount of information that must be dealt with. In consequence, humans have not needed to be fast calculators, to be permanently vigilant, or make predictable judgements to cope with their evolutionary environment. In fact, it may even be an evolutionary disadvantage to them if they did have these facilities.

The major drawback with this selective information processing in humans is that it is liable to error. While some of the errors we make are unsystematic, such as physical slips, many more arise from the way in which we interpret incoming information. Arguably, we have evolved in such a way as to tolerate a fairly high level of error because most of the errors we make are not dangerous. The vestiges of this are to be found in superstitious behaviour or avoidance of things which once caused us discomfort, such as certain unusual foods. However, in modern society, some errors have serious consequences, such that error-avoidance and recovery are major issues in designing interfaces for anything from toasters to power stations. In Chapters 2 to 5 we consider how properties of well designed interface dialogues can assist in the control of errors. The way in which skilled users learn to avoid errors is then discussed in Chapters 6 to 8.

An overview of cognitive psychology

As argued in the previous section, it is the use of mental faculties for dealing with information efficiently that underlies our evolutionary success. We do this by selecting only information that is relevant to the current task, and by using prior knowledge to understand and predict events. Mental faculties for processing information are the province of cognitive psychology, and much study has been devoted to how cognitive processes select and interpret information. It is not necessary to give a complete account of human cognitive processes to explain their importance for understanding interfaces (for a more detailed review, see Eysenck and Keane, 1990). However, in order to understand the terminology used in later chapters, an overview is presented here.

Figure 1.1 (adapted from Shiffrin and Schneider, 1977) presents a view of mental processes broadly accepted by the majority of cognitive psychologists over the past thirty years. This view reflects the concept of the human mind as an *information processor*. The details vary between different models (other examples include Broadbent, 1963; Anderson, 1983; Norman and Shallice, 1986), but the basic idea is that information comes in through the senses, is stored for a short time, decisions are made as to its relevance, processes are carried out upon it in relation to previously held information, and a response in the form of an action or utterance is made. According to this view of human mental functioning, in order that information processing may be carried out, a number of cognitive processes are necessary. These include:

sensation and perception of incoming information;
attention to relevant information;
short term storage of incoming information whilst it is processed;
memory over extended periods of useful information;
acquisition of complex skills;
linguistic comprehension of incoming information;
linguistic production of verbal or written output;
problem solving to decide appropriate actions and utterances;
planning sequences of actions and utterances.

This list represents a fairly traditional view of cognitive psychology, and a number of critiques (e.g., Neisser, 1976; Searle, 1984) and alternative views (e.g., Rumelhart and McClelland, 1986) have been offered. Even psychologists who still hold an information processing view would not now wish to equate the list of cognitive processes as discrete or serial functions, since all cognitive processes are highly interdependent. However, it serves usefully as

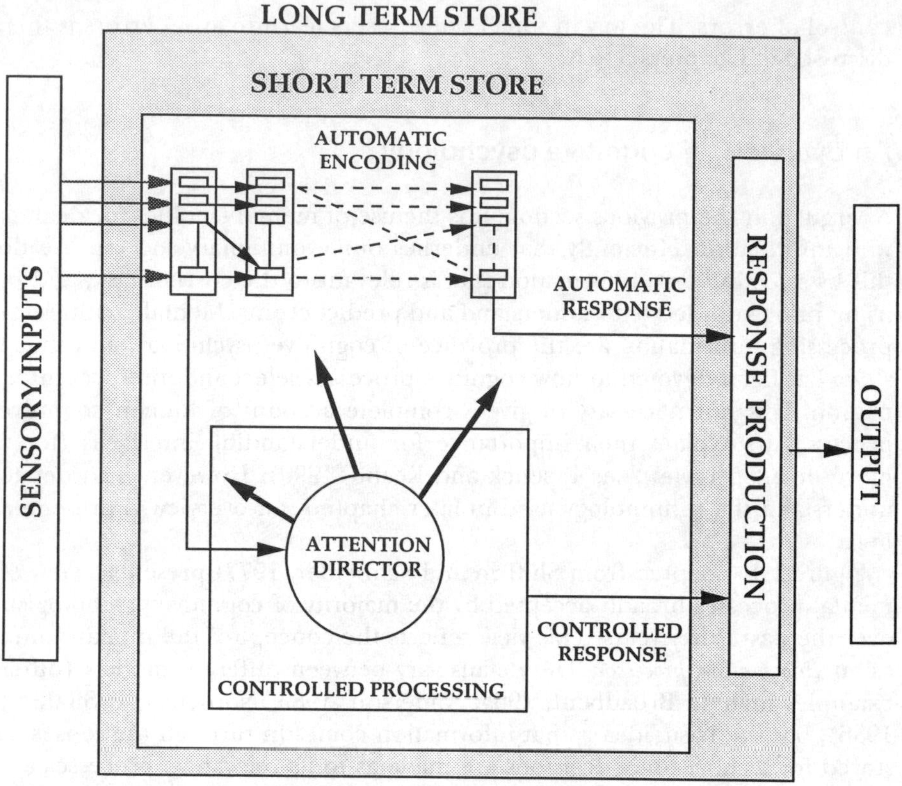

Figure 1.1 A model of the human cognitive system, adapted from Shiffrin and Schneider's (1977) theory of controlled and automatic processes in attention. The 'lightening flashes' represent a mechanism which automatically responds to incoming information and produces an appropriate action or decision sequence without requiring conscious attention. The 'attention director' is a mechanism which deals with incoming information that cannot be dealt with by an automatic process, and therefore requires conscious attention to be applied before a response sequence can be output. The arrows pointing out from the attention director represent the need to use other cognitive information sources (notably short-term and long-term memory) during controlled processing.

an approximate description of the sources of constraint and skill in interface use and design.

All of the cognitive processes listed above are constrained in one way or another. For example, our ability to process and remember information over short periods of time is restricted, both in terms of the amount of information we can hold and the complexity of the tasks we can carry out using that information (Baddeley and Hitch, 1974). Thus, whilst it is fairly easy for most people to remember a ten digit display of numbers at the same time as dialling these numbers as a sequence into a telephone, it is much harder to remember a sequence of more than, say, four or five numbers if the task is to multiply the numbers together in one's head and find a final value.

Early theories of cognitive processes attempted to quantify the constraints on cognitive processing. For example, Miller (1956) suggested that seven items (plus or minus two) was a reasonable approximation of how many items people can be expected to remember over short periods of time. The trend towards measuring constraints on human performance can be seen in some models of human-computer interaction that attempt to model the time and effort required to use different interfaces (e.g., the GOMS model of Card, Moran and Newell, 1983). However, it has proved difficult to determine exact values on the constraints of perception, attention and memory. This is because such values are never independent of the circumstances in which they were observed. For example, if you are given a list of fifteen unrelated words to remember, then the chances are you will recall approximately seven of them if asked to recall them a few seconds later. If, on the other hand, you are given a list of fifteen capital cities to remember, you will probably recall considerably more. The effects of cognitive constraints are *context-sensitive*. They depend on a large number of factors, such as the type, complexity, discriminability and familiarity of the items to be remembered and the size and nature of the task that must be undertaken with them. Unfortunately, this has not prevented some purveyors of interface design guidelines from justifying suggested features of interfaces, such as the amount of information to place on a screen display or the number of items to put in a command menu, in terms of *absolute* limits on human short term memory abilities. While the application of such guidelines sometimes result in a satisfactory outcome, they are based upon a simplistic view of human cognition and do occasionally produce undesirable results. The use of guidelines is discussed further in Chapters 9 and 10.

As we argued in the previous section, there are evolutionary reasons why cognitive processes are constrained, since human survival has depended upon our ability to select only relevant information and to ignore, filter out and forget irrelevant information. It is only when exposed to a task such as using a computer interface that these constraints can appear to take on a negative character, in that they are one of the major explanations of error in using interfaces. Thus one hears of failures to discriminate between control panel switches, failures to attend to relevant alarms, and failures to remember all the necessary information from a VDU (Visual Display Unit). As we shall see in Section 1.4, the consequences of such failures can be catastrophic.

A more positive view of the relationship between human cognitive processes and interface use is offered by considering the skilled nature of human cognition. The skills we bring to interface use include: the application of prior knowledge to understand an unfamiliar situation; the linguistic comprehension and expression of ideas; the use of abstract representations of objects for understanding and predicting events; the development of 'automatic' or expert skills; and the solution of complex problems using reasoning skills. An objective of interface design should be to allow users to express these skills whilst avoiding the imposition of activities to which the user is unsuited. Part 3 of the book focuses on the nature of users' skills and the ways in which interfaces should be designed to support them.

1.3   A broader context for understanding interfaces

Although this book is primarily concerned with psychological issues in the design of human-computer interfaces, there are many other factors that shape the way that interfaces are eventually designed. These include economic, organizational and even political factors. Of particular significance are the costs to users and manufacturers, the perceived effectiveness of new interfaces, as well as requirements of safe operation.

Costs

It is well known that people have immense difficulties in programming video recorders, yet the majority of manufacturers persevere with obtuse and difficult interfaces. When profitability is made the main factor in a cost-benefit analysis, the reason becomes apparent. Most people are prepared to tolerate a difficult, though functional, interface as long as the video recorder is cheap; until, that is, a manufacturer produces a recorder which is clearly easier to use while matching its competitors in price and performance. It is of little economic benefit to a manufacturer to replace a successful product with an improved design that offers only a minor benefit to the user. Of course, this is by no means universally true, particularly where safety is involved. However, generally a manufacturer will not rush into improving the quality of an interface unless either they gain a marketing advantage by doing so or they are required to do so by legislation.

Another reason for manufacturers to delay the introduction of improved interfaces is the high costs associated with technological development and purchase. Prior commitment to one form of technology may also prevent a change to better technologies simply because of the inertia against change and the costs therefore incurred in attempting a change. Perhaps the best known example of this is the QWERTY keyboard (e.g., Noyes, 1983) which was originally intended to *slow down* the key strokes of skilled typists in order to avoid the mechanism jamming on traditional typewriters. QWERTY remains to this day the standard keyboard pattern on modern typewriters and computers despite the existence of demonstrably faster keyboards. There are many reasons why this is so (and these will be expanded upon in Chapter 3), but perhaps the most compelling is simply that almost everyone who has used a typewriter or computer has learnt using the QWERTY keyboard. Add to this the fact that almost every typewriter and computer bought by commercial and educational establishments or by individuals has a QWERTY keyboard, and one can readily understand why alternative keyboards have not taken off. Advances in interface design are more likely to occur when they compliment existing technologies than when they challenge them.

Effectiveness

Effectiveness and cost are closely related: if one can carry out a task with fewer staff and in less time, then it is going to be more profitable. Increased effectiveness can also open new markets, as exemplified by the recent boom in direct mail marketing through the use of automated 'personalizing' of advertising material using computer databases of names and addresses.

Increased effectiveness is not guaranteed by the introduction of new technology, since it can disrupt previously efficient practices. An example of the range of symptoms and indirect costs which can afflict a company is described in a case study by Lansdale and Newman (1991). They examined the difficulties which arose when a network of confectionery distribution depots were computerized. For example, the loss of trained staff rose uncomfortably, and a number of minor errors began to occur both at the check-out tills in the depots (essentially specialized supermarkets) and in the orders department. Much worse, the warehouse managers were unable to establish the true level of their stock because of a range of difficulties with the system interface. They compensated by ordering and holding too much stock. They did this because it was the only method by which they could be sure of having the goods available to sell to their customers. It is noteworthy that their profit bonuses were based upon how much the individual warehouse sold, rather than upon the overall company profits. The consequence was that their company fell into financial difficulties as the capital invested in goods sitting in the warehouses grew too large.

The loss of efficiency in this company when they introduced a new computer system was clearly associated with both changes in work practices and poor system design. The case study also illustrates the importance of organi-

zational factors in the implementation of new systems. Although many features of the design were poor, it was the mismatch between the aims of the
company (to maximize profit and minimize stock holdings) and the aims of
warehouse managers (to maximize the amount of stock passing through
their warehouse) that gave rise to the most serious problem, which was the
over-ordering of stock.

Sometimes the causes of loss of efficiency are extremely subtle. Brown,
Wastell and Copeman (1982) compared the performance of switchboard
operators using an old plug-and-socket switchboard (generally regarded as
noisy and cumbersome equipment) with that of operators using an electronic 'cordless' switchboard where some attention had been paid to the
ergonomic design of its layout and controls. In short, the plug-and-socket
operators were more efficient than users of the new system and generally
recorded higher levels of job satisfaction. The explanation for this difference
lies in the physical properties of the old switchboard, which might not
appear to be part of the system interface. The noise of the switchroom
(caused by operators inserting and removing plugs), and the physical layout
of the web of cords on the switchboard were essential cues to the operators
as to the state of the system and the volume of calls to be handled. In the
new switchboard, this information was harder to determine. The users felt
'cocooned' from the system, they were less responsive to rapid changes in
traffic volume, and felt unhappier about their control of the task. Thus, the
relationship between efficiency and 'good' interface design is not always
straightforward. In this example, the very components of the system that
seemed particularly clumsy were also those which enabled users to carry out
their task efficiently.

Safety

In commercial system design, it is important to convince a potential
customer that productivity will be enhanced by the interface. In developing
interfaces for hazardous jobs it is essential that strict safety standards are met.
Users in these situations operate this equipment because it is their responsibility to use it, they are paid more for doing so and often enjoy the prestige
of the job. Interface design in these circumstances is strongly influenced by
the need for safety which overrides other factors such as convenience, ease of
use, and minimal training requirements. Errors are reduced by whatever
means are available, which may involve users in repetitive checking and
confirmation tasks or in lengthy training.

One response to this problem is to rely on automatic safety mechanisms
which undertake compensatory actions when abnormal or dangerous states
arise. Sometimes, this can bring with it unexpected problems. Whilst automatic safety mechanisms can handle most system problems (which occur

either through human error or machine failure), those that do require human intervention are self-selected to be the most difficult and complex problems. However, automatic mechanisms have a cocooning effect in which the operators are shielded from the system during normal operation. Thus they may not be made aware of any pre-occurring symptoms of an impending crisis. The net result of this is that, although automation minimizes the need for human input, it can also increase the seriousness of the problems which human operators are required to solve, whilst at the same time reducing their preparedness to solve them (Bainbridge, 1987).

Another response to the demands for safety is an increased reliance on training. If a complex response is required in an emergency situation, then extensive training is required so that the first action that comes into the operator's mind is the trained response. However, given the high levels of stress that accompany emergencies, it is easy to see how even the best training may fail. This is illustrated by the following example.

In 1989, forty seven passengers died when a British Midland Airways Boeing 737-400 crashed onto the M1 motorway at Kegworth. The accident occured because, in attempting to respond to engine damage in the No. 1 (left) engine, the crew shut down the No. 2 (right) engine. It is not clear exactly why the crew made this mistake. Questions have been asked, for example, as to whether the vibration dials, which functioned correctly, were adequately clear. However, the conclusion of the enquiry (Air Accident Report 4/90 Department of Transport) was that the pilots acted contrary to their training by responding too quickly to the emergency. Three explanations for this action can be given. First, on the basis of their previous flying experience in different aircraft, and their knowledge of the air conditioning system in those planes, the crew incorrectly assumed that the presence of fumes indicated damage to the right engine. Second, it so happens that abnormally high readings on the vibration meter on the Secondary Engine/Hydraulic Display correspond with near-normal readings on its near-neighbours. Cursory glances might therefore lead to the conclusion that the *right* engine was showing abnormal reading whilst the left was normal (the reverse actually being the case). Third, one of the tragic ironies of the accident was that, by turning off the right engine, the aeroplane temporarily stabilized, thus offering confirmatory evidence to the pilots that their decision was correct.

Thus, the error that led to the Kegworth air crash may have occured because the pilots made a judgement based on past experience (actually inappropriate in this case) and the misleading feedback their actions received, rather than relying on their training in emergency procedures. Although the pilots were at first commended for their bravery and skill in keeping the aeroplane airborne for long enough to avoid crashing into the village of Kegworth, they are no longer employed by the airline. Wherever the blame may lie for this accident, it illustrates the complex relationship that exists between the operator, their training, the interface, and the con-text in which operation of the interface takes place. There is always a rush to apportion blame after accidents, such as train and aeroplane crashes, in which machinery is misused or malfunctions. All too often it is the hapless user/operator (the train driver, pilot, or signal box engineer) who is held liable. We would argue that this is sometimes inappropriate. This is not to say that operators or users should never be held accountable for errors in the use of safety-critical systems, but that they cannot carry the whole responsibil-ity. People make unintentional errors: it is a fact of life, and one cannot pre-dict all the situations in which errors might be made. One can only try to design interfaces that are sensitive to the range of conditions in which they might be used. To this end, Norman (1988) argues that designers should work under a principle of designing for error, that is, they should develop systems in the realization that human errors will always be likely to occur, regardless of the training or commitment of users.

1.4   Interface design

In this section, we consider the design of interfaces such that they best match the context in which they are used. Methods for interface design are consid-ered in detail in Part 4. Figure 1.2 shows a characterization offered by Young (1983) of the principal players involved in interface design: the user; the psy-chologist; and the designer; and their relationship with the device and task to be undertaken. Young points out that a designer's conceptual understand-ing will typically focus on the device, whereas the psychologist's conceptual understanding will focus on the user (this focus being suggested in the dia-gram by the 'eyes'). Young points out that these are not the only conceptual views that might be held, and one of the recurring themes of this book is that it is insufficient for psychologists and designers to consider interfaces from a single perspective. Instead, it is necessary for psychologists and designers to hold views of interactions between the user, device and task. Thus, in Figure 1.3 we have extended Young's characterization to include a number of addi-tional views of the human-computer interface, each of which forms the basis for an understanding of interfaces that is described in later chapters.

In order to achieve a sufficient understanding of the interactions that exist between the user, device and task, it is essential that the context of interface use is analysed before any design work takes place. To do this requires that the designer undertakes a thorough *task analysis*, that is, a formal (or at least semi-formal) description of the tasks that a user will undertake with an inter-face and the requirements that each task generates for information and actions from the user and from the interface. Although many methods for task analysis exist, we concentrate in later chapters on Hierarchical Task

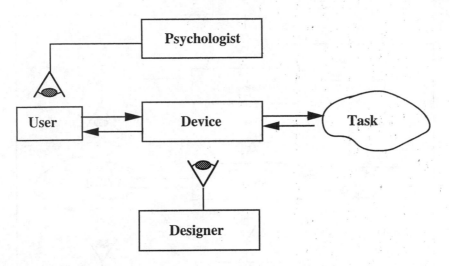

Figure 1.2 Performing a task with an interactive device (from Young, 1983).

Analysis (as outlined by Shepherd, 1985), as much as anything because it is one of the simpler methods to undertake. Of all the techniques that can assist the process of interface design, we believe that some form of task analysis is the most essential. This is not to say that other techniques, such as structured design methodologies, prototyping, psychological experimentation and cognitive modelling, are not of value to the interface designer (though we do argue in Chapter 10 against over-reliance on some methodologies). However, without having first conducted a thorough task analysis, the designer is not in a position to understand the context in which an interface will be used and develop it accordingly.

The relationship between interface design and technological change

Interface design is not simply the application of new technologies to existing tasks. There are a number of reasons why it involves more than this:

- Technology changes extremely quickly, such that some of the principles evolved for designing interfaces become obsolete. An example of this was the belief that white characters on a black VDU screen made text much clearer to read than black characters on a white background. This conclusion followed from the observation that the quality of screens were such as to blur the characters on a white background. With improvements in screen hardware, this constraint no longer applies, as anyone who has used a modern desktop computer will testify;
- There is often a loss of flexibility through over-reliance on technological

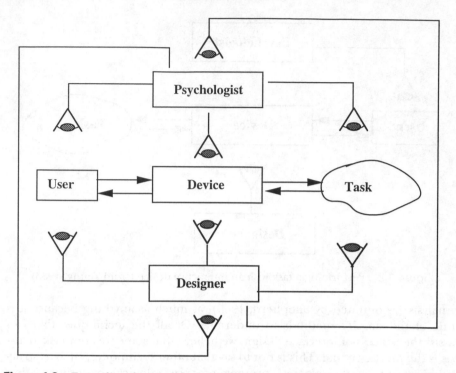

Figure 1.3   Extending the possible views of the interaction between users and devices.

solutions to interface problems. A convenient solution may create a worse problem when the circumstances of its normal use no longer apply;

- Technological change costs money. Often, money spent on technology is not always well spent, and small changes in work practice or to the task can be as effective as major innovations;

- In safety-critical systems, the reliance upon automated solutions is controversial, since they can sometimes result in a set of circumstances which present a greater crisis to the user.

The design of interfaces is not, therefore, a purely technological matter. As researchers, we need to avoid a preoccupation with studying existing technologies for their own sake. We argue that it is a thorough analysis of the task at hand, coupled with psychological studies of characteristics of the users, which will offer the most enduring information for interface designers. Provided these processes are described and understood in terms which do not depend upon technology, their implications will endure compared with the ephemera of technology.

A common description of the psychologist's role in the design of interfaces is to identify optimum characteristics for the efficient operation of a given interface. The aim of studying interfaces is seen as optimizing users' and organizations' efficiency as well as minimizing problems. The objective of interface design in this view is to achieve a 'best fit' between task, technology and user. Indeed, a recent trend has been to refer to the application of cognitive psychology to interface design as a form of 'cognitive engineering' (e.g., Norman, 1986; Dowell and Long, 1989).

We believe that this is a limited view of the potential impact of psychology upon design. Although it is common to think of computerization as the automation of pre-existing tasks, this assumes that the user's goal stays the same, even when the technology changes. As we will argue, there are a number of situations where new interfaces lead to new capabilities. For example, the development of word processing software has required changes in the skills of secretaries. Many of the skills they used with less sophisticated word processors (e.g., setting tabs to create columns) are inappropriate and must be replaced by new skills (e.g., using tables instead of tabs to format columns). Although changes such as this may initially seem obstructive, ultimately they give the user more power and increase the range of tasks that can be undertaken. Thus interface design is not simply a matter of accommodating old jobs with new devices, although this often does happen. It is also a matter of redesigning the task. Often this occurs implicitly by a gradual evolution of tasks over several generations of new interfaces. Arguably this is an inefficient method of design, and is better dealt with explicitly by tackling the question "How do we design tasks?". In the final chapters of this book, we will return to this question and offer our view of how this can be encouraged by the use of task analysis.

In consequence we argue that interface design should be seen as a modelling of the task in which the objectives of the task and the users' goal are basic consideration, and in which the technological expression of the solution is a secondary matter. This approach frees us from seeing psychology as merely reactive to new developments in technology, a pose which has effectively forced psychologists into the role of evaluators rather than designers.

1.5 Conclusions

We conclude this chapter by laying out four principles that we have referred to in this chapter and which reappear throughout the book. In effect, they represent our reason for writing this book. They are as follows:

1. Interface use is context-sensitive

Interface designers must understand the task they are designing for. Users are sensitive to apparently minor changes in circumstance, and an interface which is well designed for one task may be inappropriate for another that seems superficially similar. This is the principle of context-sensitivity. Design, therefore, requires careful analysis of the tasks for which interfaces are intended, which makes some form of task analysis a prerequisite. Additionally, the design of interfaces can also be, in effect, the design of new tasks, because different contexts and functionality provided by emerging technologies allow either altered or novel tasks to be conducted through human-computer interaction.

2.   Interface design is not simply about ease of learning

The preoccupation of interface design with ease-of-learning is a misleading theme of previous research. While not denying the obvious attraction of making interfaces easy to use, a different emphasis is sometimes needed to support the skilled user. This raises the question of whether one should de-skill a task or equip the user with additional facilities. Sometimes making an interface easy to learn may reduce the extent to which users explore the interface, thereby reducing their ability to cope with new tasks.

3.   Technology alone should not drive design

Carrying out a task is the end to which technology is merely the means. The value of new technology comes from the opportunities it offers and the constraints it imposes, and these can only be judged in terms of the tasks that users carry out. Whilst it is important to discuss different technologies, we avoid a preoccupation with the ephemeral aspects of interface technology and instead discuss how technology influences the design of tasks (and vice versa).

4.   Psychological theory can be difficult to apply to design

A distinction needs to be drawn between psychological theories which are general but can only loosely inform design and detailed theories which, although informative to design, are applicable only in highly specific circumstances. This context-sensitivity can make psychological research very difficult to apply to the design of interfaces. We aim to show how psychology can be applied to the design of interfaces. Nevertheless, it must be recognized that there is no obviously easy and effective way to apply psychological theories in practice. It is a complex activity which requires extensive knowledge of the psychological phenomena associated with different interface technologies and tasks. Thus, the rest of this book discusses how different theoretical approaches are applied, some more successfully than others, to different aspects of interface design and evaluation.

Part 2
Understanding Dialogue

In this part of the book we consider the parallel between humans communicating with other humans, and humans communicating with computer interfaces. The analogy is implicit in the common use of the term 'human-computer dialogue'. Chapter 2 looks at the relationship between the structures of different tasks and the corresponding dialogues which evolve to deal with them; Chapter 3 discusses the relationship between dialogues and the communication medium, and Chapter 4 looks in detail at the analysis of human-computer interactions in terms of their language-like elements. Finally Chapter 5 looks at how dialogues, both human-human and human-computer, are controlled on a second-to-second basis in the face of errors, misunderstandings and interruptions.

CHAPTER 2

Fitting dialogues to the task

Overview

This chapter is concerned with the ways in which human-computer dialogues can be structured to support different tasks. The main purpose of this chapter is to show how the appropriate choice of dialogue style is related to the structure of information flows in a task and the demands that the task makes upon the expertise of the user. We discuss in detail three approaches to dialogue design: question-and-answer dialogues; form-filling; and menu-based systems, and discuss how elements of different styles can appear within the same hybrid dialogue.

This discussion of fitting dialogues to tasks, like the other chapters in Part 2 of this book, is structured by considering the parallel between human-*computer* dialogue and human-*human* dialogue. To enable this, the chapter begins with an extended discussion of different perspectives on human-human dialogues.

2.1   Aspects of human-human dialogue

The Shorter Oxford English Dictionary defines *dialogue* as 'a conversation between two or more persons'. This broad definition covers a range of activities, such as the interchange between pilot and ground control as they go through pre-flight checks, cross-examination from an income tax inspector, and children telling each other jokes. How do these dialogues differ, and what do they have in common? In later sections, we examine how dialogues differ as a function of their purpose. First, we are concerned with understanding the term 'dialogue' better by considering different aspects of human-human dialogues and the way they are influenced by other factors. We break 'dialogue' down into five factors, which we illustrate in Figure 2.1. The figure also illustrates the analogy between human-human dialogue and human-computer dialogue, and notes the chapters in which the relevant factors are covered.

Human-human communication

Human-computer interaction

Figure 2.1 A definition of 'dialogue' in terms of the comparable elements of human-human communication and human-computer communication. The examples given are discussed in the relevant chapters.

Linguistic competence

To conduct a dialogue necessitates some degree of linguistic competence. Each of us has the ability to hold up our end of the conversation in terms of constructing intelligible sentences and understanding other people's speech. At this level of analysis, two types of knowledge are relevant: a vocabulary of the meaning of words and concepts; and a grammar which describes how the meaning of utterances is contained in, and extracted from, the sequence in which words are combined.

Human-computer dialogues require a vocabulary that associates labels with concepts or actions, such as command words (words which get the computer to do something), particular materials (for example, the word *gunge* has a precise meaning in the domain of process control), and categorizations in a filing system (e.g., the *pending* file contains all unprocessed documents). The characteristics of labels in interface design are that they normally refer to specific objects or actions and are often jargonistic. The sequences of actions and commands required by some interface dialogues can be compared with the grammar that structures natural language. These rules of order and syntax carry meaning. For example, in English *John hits Peter* means something different to *Peter hits John*, and in the same way, in MS-DOS the command *COPY file1 file2* is different to *COPY file2 file1*, which reverses the action. Similar sequential dependencies hold true for other tasks, such as the order in which arithmetic functions are executed using calculators (Young, 1981).

Controversy exists as to whether the uniqueness of human language means that we have an innate grammatical skill (Chomsky, 1965; see also Anderson, 1990, for a recent review). The complexity of this debate is beyond the scope of this book, but it is interesting to wonder whether, as has been suggested by Green (1980), the cognitive processes underlying grammatical analysis overlap substantially with the understanding of artificial dialogue styles such as programming languages. One can draw an analogy between rules that link actions with the tasks they implement at an interface and the rules of grammar that determine how words are combined to make meaningful sentences. In some theories of human-computer interaction this analogy is made explicit, for example in Payne and Green's (1986) Task-Action Grammar analysis of interface dialogues. The issues of interface vocabularies and grammars are discussed in Chapter 4.

Conversational competence

Human dialogue is more than a question of grammar and vocabulary. The correct response to the question "Could you pass me the salt" is probably not simply to say 'Yes'. Behind our knowledge of how to behave in such circumstances is a pragmatic skill which is based upon the cooperation necessary for successful conversation (Grice, 1967). Conversational competence can be seen as the 'glue' which holds dialogues together and brings them back together when they go astray. The ease with which human-human dialogue normally proceeds (compared with many human-computer dialogues) has made discourse and conversation topics of interest in the design of interfaces (e.g., Luff, Gilbert and Frohlich, 1990).

A range of psychological mechanisms can be identified in human-human dialogue that lend coherence to conversation. One mechanism is the ability to infer the dialogue partner's intentions in order to make an appropriate response to an utterance which may lack sufficient vocabulary and grammatical information to be understood. Another mechanism is the *repair* of dialogues, in which breakdowns in understanding are corrected by a sequence of utterances whose intention is to focus upon the misunderstanding and put it right, as in this example:

Teacher (to parent):   "Your boys are revolting."

Parent: "Well! The others are no angels either."

Teacher: "Oh no - I meant they have realized they don't
                       have to go to ballroom dancing lessons if they
                       don't want to."

The maintenance of human-computer dialogues also requires an ability to infer the *intentions* of the dialogue partner, be it human or machine. Research into artificial intelligence has not yet provided generalizable systems that can reliably infer intentions from under-specified input from users. Therefore, interfaces must be designed in such a way that users impart their intentions in an unambiguous fashion. This means that users either have to be given a range of distinct options, from which to choose a desired input, each of which the machine can 'understand', or they must learn an artificial language that has been designed to be unambiguous. Whilst errors in human-human dialogue are easily spotted and corrected, the recovery from errors can be so difficult in human-computer interaction that users become overly cautious. These issues are covered in Chapter 5.

Nonverbal conversational skills

Analyses of human-human dialogues show a range of nonverbal cues which can be interpreted as part of the conversation. These include gestures for emphasis, gestures which disambiguate, eye movements and other kinds of 'body language'. Nonverbal skills emphasize turn-taking (body and intonation cues which mean "I have finished - now it's your turn"); they indicate a difference between a statement and a question; and they can indicate when a misunderstanding or mistake has happened. Nonverbal skills add coherence to a dialogue and can provide parallel information to the spoken word.

One implication of nonverbal skills for the design of interfaces is that many communication channels are worthy of exploration for human-computer dialogues. Methods already in common use include hand movements (mouse) and auditory feedback. There is also ongoing research into the use of eye movements (Jacob, 1990; Starker and Bolt, 1990), facial expressions (Takeuchi and Nagao, 1993) and foot movements (Pearson and Weiser, 1988).

In order to use different channels for human-computer interaction, we need to differentiate between the *intention* of a communication, and the *medium* through which it is carried out, since this limits the range of gestures that are possible. For example, one might expect telephone conversations to be harder to conduct than face-to-face dialogues because the lack of visual information makes them *cueless* (Rutter, 1987). In practice, there is no evidence for systematic miscommunications in telephone conversations (Drummond and Hopper, 1990) because other means are found to manage the dialogue, one method being a reduction in the number of attempted interruptions. Thus, when some communication channels are closed, alternative methods are found to achieve the purpose of the conversation.

Generally, humans are good at adapting their conversation to ensure that its overall purpose is achieved. However, some cases of miscommunication come from speakers adopting inappropriate strategies. For example, Coleman and DePaulo (1991) found that people often adopt a form of 'baby talk' when conversing with physically challenged people. Maladaptive strategies such as this represent an inability to understand the speaker's point of view, or the application of an inappropriate belief based upon prejudices or prior assumptions. A similar failure to form a shared understanding can be seen in some forms of human-computer interaction. There is a tendency, with machines which show any form of sophisticated linguistic interpretation, for users to assume they are capable of much more than they actually are. For example, novices often develop misconceptions when learning the programming language Prolog, because they assume that the system 'understands' the programs that they write, since simple programs can bear a passing resemblance to English sentences (Taylor and du Boulay, 1986; see also Fitter, 1979, and Suchman, 1987, for additional examples of inappropriate models of shared understanding).

Constraints of the communication medium

Human-human communication is affected by the medium through which it travels. For example, Chapanis, Ochsman, Parrish and Weeks (1972) found that pairs of subjects who solved problems through communicating by sound only exchanged ten times the number of messages compared with those who communicated by writing. This presumably reflects the inconvenience of writing and the consequent need for economy of expression. In general, the properties of communication channels force a number of modifications on user behaviour, such as slowing down, avoiding ambiguous utterances, and requiring confirmation of receiving a message. In the design of interfaces, input devices (keyboards, mouse, touch screens) and output devices (screen, auditory feedback) mediate between system and user, and therefore influence the dialogue that takes place. The role of the medium is discussed further in Chapter 3.

Task constraints

The *task* (possibly in conjunction with medium constraints) can also determine the structure of human-human dialogues. For example, the language between pilots and air traffic controllers consists of short utterances which focus upon communicating specific information (heights, bearings, and landing and take-off position), interspersed with call signs and repetitions. This restricted form of dialogue is designed to maintain safety and efficiency in a crowded air space. Other examples of formalized dialogues are to be found in situations like military communications, legal proceedings, and in dentistry. In these cases, to avoid misunderstandings or to communicate complex information accurately, the constraints of the task drive the dialogue away from 'natural' conversation into restricted constructions which are less prone to misinterpretation. Arguably, notations such as music and mathematics are types of formalism to communicate highly specialized information.

Human-computer dialogues can be seen to reflect the need for a task focus. Making them as 'natural' as possible is not purely a technical matter of designing machines which have the necessary facilities for comprehending human language. Indeed, for certain tasks it may be desirable to design dialogue styles which support restricted styles of dialogue which are deliberately unlike the natural language of conversations.

2.2   Dialogues and interface tasks

The tasks for which we use machines vary in complexity from something we can all do, such as getting a ticket from a vending machine, to skilled operations, such as controlling a power station or flying a plane. Often, the human-human dialogues which sustain such tasks are adapted to meet the demands of the task. This is most obvious in specialized and skilled tasks, such as air traffic control or military communications. The restricted language which supports these tasks is sometimes referred to as an *operative language*. Such languages are characterized by a restricted vocabulary, domain specificity, and a use of grammar which is economical to the point of distortion (Falzon, 1990). Some tasks are inherently complex, exacting or dangerous, and what is a 'natural' dialogue style for these tasks may not apply to others. Using human-human dialogue as a model, we should not be surprised to find that dialogues for human-computer interaction can also differ according to the tasks they support. Thus, a number of dialogue styles have been developed to support different interface tasks.

Another significant factor in maintenance of human-human dialogue appears to be the expertise of the participants. For example, Falzon (1990) describes dialogues between experts and non-experts (as defined by their knowledge of the subject matter) in situations such as patient-doctor interviews and telephone-based technical support. In these circumstances, the expert speaker assumes control of the conversation, usually as soon as socially acceptable, and the remaining exchange often follows a sequence of questions and answers, usually of the 'yes or no' variety. This contrasts with the situation where an expert speaker perceives that the other speaker is also skilled in the domain, in which case they are less likely to try to control the conversation and so conversation is more focused and technical.

What aspect of expertise (or lack of it) necessitates that dialogues follow these different patterns? One important factor in the maintenance of such dialogues is the speaker's prior knowledge and its effect upon his or her recall. By definition, non-experts know little about the tasks they are involved in. When, for example, you consult with your lawyers, they take responsibility for making sure that all the necessary procedures are followed and provide you with questions or simple requests for information. Here the expert knowledge is one of procedure, and the reason for the lawyer taking the initiative is to ensure that nothing is forgotten or overlooked. As an example of this in human-computer interfaces, automatic bank machines ensure that the customer walks away with both cash *and* card by imposing a procedure in which the card, which could easily be forgotten, is removed before the cash can be removed. Many human-machine dialogues are therefore structured to ensure that users follow a given procedure when they cannot be assumed to carry it out reliably. Only skilled individuals can be expected to remember complex procedures. In these circumstances a less structured, dialogue may be possible.

Studies of expertise have consistently shown that the ability to assimilate and process information depends upon how much is already known about the structure of that information (see Anderson, 1990, for a review; expert performance is discussed further in Part 3). This is because prior knowledge allows the expert to categorize and compress incoming information into known patterns which appear meaningless to a novice. It follows that experts will find it much easier to keep track of where they are in a complex task and to make sense of the information they are receiving.

An example of expert performance in comprehending a dialogue can be found in the domain of music. Figure 2.2 shows part of a Beethoven piano sonata. If you cannot read music, you might be surprised to learn that experienced musicians are able to read and play this music at sight. They are also able

Figure 2.2 The first page of Beethoven's Waldstein Piano Sonata. Those who can not read music are surprised by how much musicians can remember of this complex array after being shown it for only a short time.

to reproduce this page, and more, exactly as it is written after just a few seconds viewing. They can do this both because they may know the sound of the music and can therefore reproduce it from memory, and also because it is full of patterns of music which are relatively easy to replicate once they are recognized.

Reading music and playing musical instruments is not so different from human-computer interaction as might first appear, since instruments *are* machines and music is an example of a *notation* for encoding instructions. In this case, it is a notation which instructs the musicians as to which notes they should play, and when and for how long they should play them. Like jargon, such notation has evolved because natural language is not able to communicate the complexities of the composer's intentions. Experts invest effort in learning notations and jargon precisely because of the precision with which they communicate information. They use shorthand, jargon and operative languages, not just because it sounds impressive (though sometimes one has one's suspicions!), but because it allows for more efficient communication of complex ideas.

Human-machine dialogues reflect this need for precision with different *styles*. Some (notably command languages such as UNIX) offer great power and flexibility to the user if they are prepared to learn how to use them, while others are easier for the non-expert to use because they make fewer assumptions about what the user knows and can keep track of. We discuss command language dialogues in detail in Chapter 4. These are treated separately, because unlike the dialogue styles to be discussed below, the dialogue is entirely user-driven. Many dialogues give minimal feedback as to what is going on inside the computer, with the dialogue driven predominantly by the user, as in this example of a segment of dialogue from the command language of the MSDOS operating system:

> del b:*.*   {the user deletes all files from drive b}

> copy c:*.rec   {the user copies all files suffixed 'rec' from

                     drive c to the current directory}

> SNOBOL4 sprog   {the user runs a program written in Snobol4
                     called sprog}

An important feature of this dialogue is the absence of machine-imposed structure on the sequence in which the user can type command lines. The user is not constrained to particular sequences of commands (although the internal syntax of those commands may be complex and constrained) and may well have several hundreds of operations available. This style of dialogue contrasts strongly with the three styles of dialogue on which we focus in the rest of this chapter: question-and answer, form-filling, and menus. Few interfaces can be described purely in terms of one style or another. For example, the style of interface for computers used increasingly in modern personal computers (such as the Microsoft Windows and Apple Macintosh interfaces), the so-called WIMP interface (Windows, Icons, Menus and Pointers) is a hybrid, with all these styles being employed for different aspects of the interface. This reflects the fact that, in designing interfaces for complex tasks, different styles are appropriate to different tasks faced by users.

2.3   Question-and-answer dialogues

Just about the least demanding task, in terms of the requirement for cognitive skills, is to do as you are told. Many machine-led dialogue styles prompt the user with a demand or a question and require a simple response, such as this exchange at a cashpoint:

```
Enter amount requested and press PROCEED to continue
```

Such exchanges are referred to as question-and-answer dialogues because they often takes the form in which the machine prompts with questions and the user answers:

```
Do you wish to continue? (press Y or N)
```

Machine-led dialogues have some clear advantages and disadvantages by virtue of the way they proceduralize tasks into a sequence of instructions. This works well for tasks which naturally fall into stages because the user is led infallibly through the necessary sequence of events, doing no more than making responses to machine requests. The opportunities for losing the thread of the interaction, or being unable to complete the task, are minimized. For this reason, question-and answer dialogues are usually considered suitable for inexperienced or occasional users, or for infrequent tasks.

When are question-and-answer dialogues appropriate?

There is a received wisdom that the simpler the dialogue style, and the more initiative that is taken by the computer in driving the dialogue, the better it is for novice users and the worse for experts. This carries some face validity in that novices evidently struggle to use complex interfaces and experts often resent being required to wade through a number of simple questions where a single complex command statement would do instead. However, the balance of evidence is that this is not a useful rule for interface designers to follow. As we shall see in Section 2.5 when we discuss menu-based dialogues, the prediction that novice users prefer simpler menu systems has been shown to be wrong in some experiments (e.g., Whiteside, Jones, Levy and Wixon, 1985). This creed reflects a confusion between *style* and *information content* of a dialogue. If experts *do* become irritated by question-and-answer dialogues, it might be because they are being asked the wrong questions: a different sequence might be devised that is perfectly acceptable to achieve a task given their level of expertise.

Instead of basing a decision on whether to use question-and-answer dialogues solely on the level of expertise of the intended users, we argue that it is more important to consider the nature of the task for which the dialogue will be used. Certain tasks suit a question-and-answer dialogue whatever the user's expertise, particularly when they involve the elicitation of discrete bits of information. If the user's task can be effectively carried out by answering a pertinent sequence of questions (as judged by the user), then the level of expertise of the user is irrelevant. For example, automated cash dispensers employ question-and-answer dialogues effectively, because the user only need provide one bit of information for each decision that the system must make. Some expert systems (i.e., computer-based systems which use a database of knowledge and a reasoning mechanism to simulate the judgements made by domain experts) operate with a question-and-answer dialogue because this is the most efficient way of eliciting information.

The use of question-and-answer dialogues breaks down if the user must supply large amounts of information for each function that the machine carries out. Some tasks do not lend themselves comfortably to such rigid sequences of action without becoming rather tortuous and adding difficulties to the interaction. Consider, for example, a stock control task in which the user has to enter different product line quantities for a number of months, as shown in Figure 2.3a. A question-and-answer dialogue to construct such as set of quantities is shown in Figure 2.3b.

This is rather tedious, and it is easy to see how errors in such a dialogue might arise simply by not paying sufficient attention or losing track of which question relates to which cell on the display. Sometimes previously answered questions scroll off the top of the screen in such a way as to place a memory load on the user when the answer to the current question depends upon what was given earlier. Suppose, in the above example, the same sequence of questions might have to be asked several times, one for each warehouse in a chain. Then the above questions would have been preceded by an earlier question of the kind:

```
Enter the depot name    COVENTRY
```

It is not hard to see that if this information disappears, then the user can lose track of which set of stock control data they are entering.

2.4   Form-filling dialogues

Repetitive tasks, as well as tasks which require the user to input large amounts of data before a machine action can be undertaken, are best served

Depot: COVENTRY				
	May	June	July	August
Component				
RB13	231	431	76	230
RB14	132	46	98	133
RB15	324	10	52	54
RB16	657	3	81	102

Figure 2.3a   Simulated data in a stock control exercise.

by *form-filling dialogues*. These consist of displays resembling paper-based forms in which the user enters data in specified fields, as exemplified in Figure 2.4. There are three main advantages to this style of machine-led dialogue when the task involves the entry of related data. First, the compatibility with paper-based information is preserved. Structures by which information is represented on paper (e.g., an accountant's ledger) have not evolved by accident, but serve important functions for organizing and highlighting

System Questions	User's responses
Enter Depot name:	COVENTRY
Enter rb13 units for May	231
Enter rb13 units for June	431
Enter rb13 units for July	76
Enter rb14 units for May	132
Enter rb14 units for June	46
Enter rb14 units for July	98

Figure 2.3b   An example of a question-and-answer dialogue to enter the data shown in Figure 2.3a. Note that question-and-answer dialogue is not always expressed as questions: in this case the system requests for information are in the form of an instruction. The effect upon the user is the same as if questions had been asked.

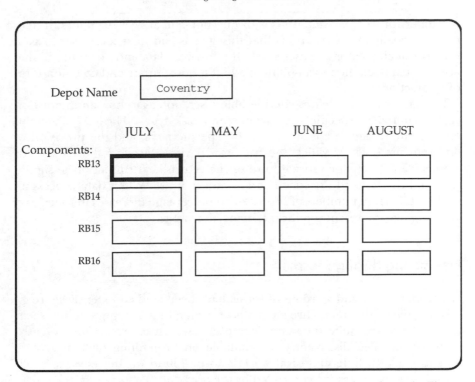

Depot Name   Coventry

JULY   MAY   JUNE   AUGUST

Components:
RB13

RB14

RB15

RB16

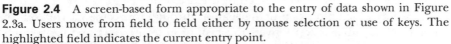

Figure 2.4 A screen-based form appropriate to the entry of data shown in Figure 2.3a. Users move from field to field either by mouse selection or use of keys. The highlighted field indicates the current entry point.

relationships that exist between the groups of information. Second, the user is reminded about the items that should be supplied in the dialogue by the fields shown on the form. Third, the presence of related data on the same screen is useful in tasks where cells have to be compared, a feature that is lost in question-and-answer dialogues.

Unlike dialogue styles such as question-and-answer and command languages, form-filling has no obvious conversational analogue. This presumably reflects the specific nature of the information exchange that occurs with form-filling. Forms support integrative tasks in that related data is handled within a given format on the same display. Conversation, being sequential and transitory, cannot keep such information available without placing a memory load upon the speakers. Forms act as a kind of external memory, that is, a way of representing information externally which lowers the demands on the user's memory. In this way they facilitate the management of tasks which require the communication of large amounts of related information.

Like question-and-answer dialogues, form-filling is seen as a useful style for novice users because the form structures the user's task and acts as a reminder for what is wanted. As with question-and-answer dialogues, it does not necessarily follow from this that this style is bad for expert users. Again the critical element is the user's task. If it involves data entry or retrieval that uses related data, then form-filling is a highly appropriate and common form of interaction.

In fact, form-filling leaves considerable discretion as to how an interaction can be tuned to specific user groups and task types. The use of default entries, the design of specific forms for specific tasks, and the provision of different methods by which the user can navigate around the data fields (mouse, arrow keys or repeated use of the return key) can all make significant contributions to the speed and accuracy of use. When data processing staff handle large numbers of forms, small percentage increases in speed and accuracy may represent significant financial benefits.

Navigating through forms

Whilst the layout and wording of forms have been studied extensively, (e.g., Wright, 1984), the interactive elements of form-filling dialogues do not seem to have attracted quite the same degree of research interest. Consequently, although detailed discussions are available on form-filling dialogues (e.g., Sutcliffe, 1988), little empirical work is to be found on the ways in which form-filling dialogues are controlled by the user.

One study which did examine navigation through forms was carried out by Fitch (1984). Subjects were given the task of entering data about products into a database using two different methods for form navigation. In one version, subjects navigated about the form by use of the RETURN key – every time it was pressed, the cursor moved to the next field. In the other version, the users moved from field to field by use of ARROW keys which controlled the direction (up, down, left and right) in which the cursor could move. When forms contain both rows and columns of fields, the two navigation methods differ in terms of their perceived efficiency. The RETURN key method is simplest to execute, but if few inputs are required per form, it requires several presses of the RETURN key to move between each entry field while skipping fields that are not used. The ARROW key method is conceptually more complex, requiring the use of four keys rather than one, but it allows the user to navigate directly to data entry fields passing through fewer redundant fields. In the experiment, subjects' preferred navigation method was taken as a measure, and the number of data items to be entered per form was varied from one to thirteen. In addition, two response times were used for the RETURN key method, either slow (about 780 ms) or fast (about 140 ms).

The results of Fitch's experiment are shown in Figure 2.5. They illustrate an apparent trade-off in terms of users' preferred navigation method

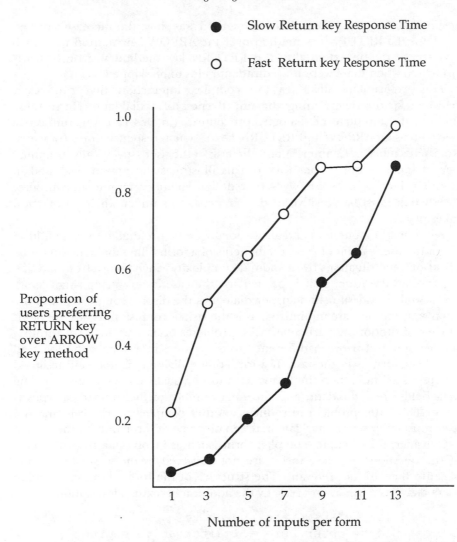

● Slow Return key Response Time

○ Fast Return key Response Time

Proportion of
users preferring
RETURN key
over ARROW
key method

Number of inputs per form

Figure 2.5 User preference of Return-Key over Arrow-Key methods for navigating
form fields. Subjects' tasks varied as a function of the numbers of data items to be
input per form (from 1 to 13) and the delay built into the system response time in the
Return-Key method (from Fitch, 1984).

(RETURN vs. ARROW keys) which was affected by the response time of the
Return method (slow or fast) and the number of data inputs required. The
proportion of users preferring the RETURN key method over the ARROW
key method increased sharply as the number of data items to be entered
increased. This reflects the fact that the RETURN key method is most efficient when there are fewer redundant fields to skip. However, when the

response time of the RETURN key method was slow, the increased preference for the RETURN key method over the ARROW key method was much
less sharp, to the extent that the RETURN key method was only clearly
preferred when there were no redundant fields to be skipped.

This experiment illustrates the complex interaction that can occur
between factors determining the effectiveness of a dialogue. Here, these
include the minutiae of dialogue procedures (in this case the difference
between using ARROW and RETURN keys), system response times (of which
we discuss more in Chapter 5) and the task of the user (specifically, the number of inputs required per form). This illustrates the notion discussed in
Chapter 1 of *context-sensitivity* – that design solutions in human-computer
interaction must be sensitive to the circumstances under which interaction
takes place.

When might a question-and-answer dialogue be preferable to form-filling?
From the user's point of view, the benefits of a form-filling dialogue might be
offset in some situations by the added complexity which comes from navigating around the form. This is particularly the case in answering single questions, while in question-and-answer dialogues the display complexity and the
details of response are minimized, as is the risk of confusion. From the point
of view of supporting particular tasks, problems of navigation, both within a
single form and between different forms, become more acute the less well
suited the form is to the task. If a form-filling dialogue is not well designed
for the users' task, then the forms tend to be visually complex, many of the
data fields are redundant, and the total number of different form types is
large. This latter problem introduces another complexity, which is that the
user must learn how to navigate between one type of form and another.

Consider, for example, complex forms such as income tax returns. Many
of the questions on these forms are not answered by the majority of users
because they are not relevant. The structure of the form attempts to guide
users away from irrelevant areas, by the addition of textual descriptions:

```
Q. 17: If the answer to the above question 16.1a was
'yes' please enter here all alternative sources of
income other than those recorded in question 14.
```

Unfortunately, the general-purpose design of income tax returns creates
interdependencies between the information contained within previously
completed fields and current or later fields (such as the complex relationship between the contents of fields 14, 16 and 17 in the above illustration).
Navigation between fields is therefore complicated, since it requires that
users navigate backwards as well as forwards in completing the form. These
kinds of interdependencies can be avoided in question-and-answer dialogues,
in which subsequent questions can be tailored to be contingent upon earlier
answers. The horrors of official forms like tax returns have long been a subject of psychological research (e.g., Wason, 1961), and improvements have

been slow in coming. The danger for interface designers is that, in computerizing existing paper-based forms, they may compound the problems of poor form design by creating a navigational burden that is much greater than that faced by users of equivalent paper-based forms.

2.5 Menu dialogues

Perhaps the most common form of machine-led dialogue style is the *menu*. The user is given a range of selections from which the required choices are made, similar to the way in which they choose from a restaurant menu. Options are laid out on the screen display, as illustrated in Figure 2.6, and the user selects from one of them. As with question-and-answer dialogues, the user's response is to make a choice, though generally from more than

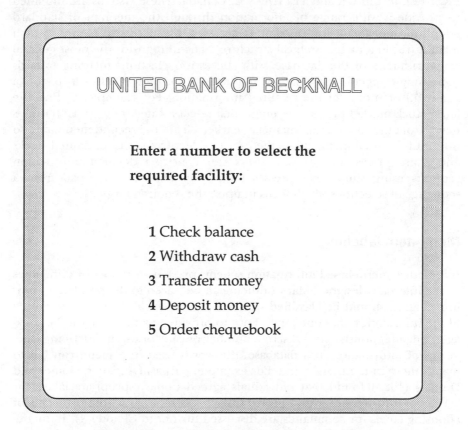

Figure 2.6   A typical menu screen for a cashpoint.

two alternatives. Again, interactions involve a sequence of selections from system-prompted alternatives, though in the case of menus the branching at each decision point is more flexible, with as many options as there are menu items.

Whilst systems that have question-and-answer and form-filling dialogues usually prompt the user to provide information, in menu-based systems the user's task is primarily one of selection from a fixed set of alternatives. Menu systems are most valuable when the users do not need to know, or cannot be expected to know beforehand, what the valid range of inputs to the dialogue are. Two tasks in which menus are commonly employed because users may not already know the required inputs are information retrieval (for example, on-line library catalogue systems) and command issuing (for example, graphics-based operating systems).

Since the mid 1970s there has been extensive research into identifying the optimum menu structures to be used in information retrieval systems. The impetus to this was the introduction of information retrieval systems such as PRESTEL in the UK and TELIDON in Canada. These systems are intended to provide a wide range of information through the medium of standard telephone networks. Information is accessed by successive menus (which form a tree-like or hierarchical structure), beginning with the most general categorizations of the database, with successive selections focusing towards increasingly specific information. Ultimately, the user accesses a 'page' (a screenful or more) of the required information. For example, to find the latest stock market prices, one might first select a category such as 'business news' from the first menu, 'financial markets' from the second menu, and so on, until the final menu allows the user to select from a list of named stocks and share prices. In this way, users can construct detailed information requests using simple key presses on an alphanumeric key pad from a sequence of selections which focus in upon the required target.

Menu item labelling

In practice, menu-based information systems present a number of difficulties to the interface designer. Many of them can be traced to the problem of how information should be classified within an hierarchical database. Very little of the information that one might want to provide in a database can be packaged unambiguously into a discrete number of categories. In describing categories of information in a database, the words used in a menu can fail to match those in the users' mind. For example, Furnas, Landauer, Gomez and Dumais (1983) found that individuals agreed upon appropriate labels for classes of information with a probability of less than 20% (the problems of choosing labels for commands are discussed further in Chapter 4). In menu-based dialogues, this problem has the side effect that incorrect choices can result in users 'getting lost' in systems (Lee, Whalen, McEwen and Latremouille, 1984). For example, suppose you are seeking information on medical insurance, does one search the database under 'insurance' or 'medicine'? When using a menu system in which one choice leads to another, an incorrect choice may go undetected until such time as the user notices that the choices offered by the system are diverging from, rather than converging upon, the required topic. In this sense, menu-based databases are rather like mazes: a turn in the wrong direction and you can get lost. Indeed, it has been found that errors can occur in up to 50% of menu-driven interactions (Bush and Williams, 1978).

Other sources of error in the use of menus are more subtle and can be harder to anticipate in design. For example, Young and Hull (1982) found that many errors are due to users failing to understand the structure of the menu as a whole. Their analysis of errors in the use of PRESTEL suggests that users perceive menus according to what they expect them to contain. Consider, for example, the user of a tourist information system searching for information about the Tower of London who is faced with the screen shown in Figure 2.7. The place names offered in the menu overlap since the London area is given as a discrete category (because it represents a large quantity of the data). The user may treat the menu items as mutually exclusive categories (which is implied by such a list), and may be forgiven for scanning the list as far as England and making this choice. Clearly, any list of options which has an unbalanced internal structure (in this case one in which a sub-category is given equal status with its super-ordinate category) is more likely to produce errors than one in which the categories are exclusive.

To summarize, the difficulty of navigating around a tree-structured database is increased if the choices offered to the user are unclear, obscure or structured in a way that differs from the user's expectations. Poor labelling and structuring makes the selection of each individual menu harder and increases the likelihood of error. Many of these problems can be ironed out by evaluating and subsequently modifying troublesome menus. However, one might doubt whether all such difficulties can be eradicated, simply because individuals differ in their interpretation of label meanings (Furnas *et al.*, 1983). One must always expect some users of some menus to fail to understand the meaning of certain categories in the way intended.

Depth versus breadth in menu design

The design of tree-structured databases allows for a trade-off between *breadth*, the number of choices allowed per menu, and *depth*, the number of menus to be traversed before arriving at the required information. Achieving the correct balance between breadth and depth can be a difficult problem for

Figure 2.7 (Top) An example of a menu whose internal structure might lead to errors. (Bottom) Care in the order of menu items, with the subset category shown first and distinguished by typeface, can minimize potential problems.

interface designers (see, for example, Paap and Roske-Hofstrand, 1988). The fewer items there are in each menu, the less specific a user's choices can be, and the more menus that must be negotiated. The more menus there are, the greater the 'distance' that must be navigated by the user in terms of the number of intermediate menus between starting point and target. Given the problems of navigating around databases, this increases the opportunities for error. On the other hand, the more items there are, the more there is to search, and the more difficult it will be to create menu item labels that are distinctive (see, for example, our description in Chapter 4, Section 4.2, of confusions that occur between the Hypercard navigation commands *back* and *previous*).

No firm recommendation as to relative breadth and depth of menus can be drawn, since the task might predispose the designer towards either breadth or depth in the design of menu-based systems. Broadly, the argument for greater *breadth* seems to be that users commonly experience difficulty navigating between menus, and specifically, getting back to where they came from once a false lead has been detected. One of the arguments for greater *depth* is the problem of crowding, where the increase in menu items which accompanies greater breadth creates problems of display clarity. Some of the errors observed by Young and Hull (1982) may be attributed to effects of crowding, since the greater the number of choices, the less clear is the relationship between them.

Despite the complexity of this trade-off, some useful generalities can be drawn. For example, MacGregor and Lee (1987) noted that the search time of a menu system of p levels is given by the following function:

$$\text{Search Time} = p\{E(A)\,t + k + c\}$$

where t is the time to read a single option, k is the time to make a selection, c is the computer response time for moving onto the next menu, and $E(A)$ is the number of options that a user examines in each menu before making a choice. For an information system containing n documents using p levels and a options per menu, the following relationship holds:

$$\ln n = p(\ln a)$$

Hence, a neat relationship can be drawn between breadth (represented here by the parameter a) and depth (parameter p) in terms of the search times, because the second equation can be substituted into the first equation to give:

$$\text{Search Time} = (\ln n)\{E(A)\,t + k + c\}/\ln a$$

which relates search time to the breadth of each menu given a fixed database of n documents to be accessed. MacGregor and Lee show that search time is minimized for menus containing between 4 to 8 items, although faster readers are considerably less affected by breadth than are slower readers. In the design of a menu hierarchy, the use of such methods of analysis are a useful source of insight, especially when one considers the appropriateness of some of the assumptions. For example, in frequently traversed menus, users may become familiar with the structure and location of individual items. As a result, E(A) decreases because the emphasis on search is reduced (the user knows where to look), and so the system can tolerate broader menus without jeopardizing overall search times.

Interestingly, improvements in screen quality and interface design may have made the problem of menu crowding less acute. For example, the *paste function* menu in Microsoft EXCEL spreadsheets, illustrated in Figure 2.8, offers a menu of some 150 different functions. It does this by showing the items in a scrolling window (or 'list box') which occupies a relatively small area of the screen and moves up and down the menu as required. The number of functions that it contains is well outside the number of menu items recommended in guidelines, which generally suggest no more than 15 separate items per menu (e.g., Williges and Williges, 1984).

There seem to be at least two reasons why very long menus can work. First, the tasks to which these menus are applied lend themselves to menus of considerable breadth and no depth. Control menus, such as the *paste function* menu referred to in Figure 2.8, present a list of items which are all discrete examples of the same category. Other examples are lists of fonts and styles. In such menus, no further logical nesting of the concepts can be achieved.

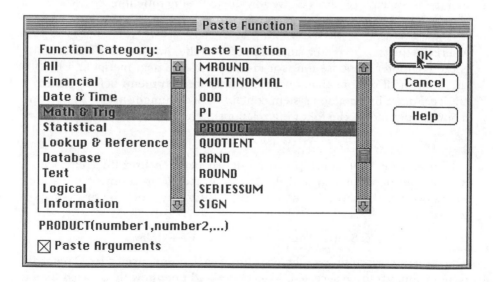

Figure 2.8   Scrolling menus used for selecting cell functions in the Microsoft Excel spreadsheet system.

Conceptually, the user's task is to know which item they want and the menu label that describes it. If the menu is long, they may additionally need to know how it is structured (for example, alphabetically) in order to facilitate the search. In this case, there are few opportunities to 'get lost' in the menu system.

The second reason why longer menus are becoming more common is the proliferation of interface mechanisms for handling menus. Menus can be implemented in many different ways, including: *pull-down* menus, in which a row of menu items at the top of the screen can be activated to reveal a menu; *pop-up* menus, in which a menu can be produced at the current point of the cursor; and *peel-off* menus, in which a menu can be moved to a position of greater convenience. Some menus have scroll bars (as illustrated in Figure 2.8), while others retain histories of how they have been used in the recent interactions to enable users to access a document or data set directly without needing to navigate through a complex hierarchical filing system. For a comparative evaluation of different kinds of menu see Macleod and Tillson (1990).

All of these facilities have become available as additional computing power has been added to interfaces, and presumably they have arisen to meet particular task demands or to enable faster working. Most of these techniques were not available to the designers of systems like PRESTEL and TELIDON some 15 years ago, and nor are they still, since the constraint of telecommunications speeds limit the responsiveness of the interface upon which these devices rely. Thus there is a marked contrast between some command and database interfaces in the ways in which menu systems are designed.

Perhaps one of the most interesting aspects of menus is how little users seem to remember about them despite the efficiency of their use. Mayes, Draper, McGregor and Oatley (1988) found that even experienced users failed to *recall* (as opposed to recognize) the menus used in the MacWrite word processing system. On the other hand, Card (1982) found that, as users learn to use a menu system, their knowledge of the locations of required commands increases, to the point where they can find the target with their first eye fixation. It appears that experienced users learn only enough about menus to enable them to search them effectively when they need to use them. This is an example of what Larkin (1989) calls *display-based problem solving*, a theory of human cognitive skills which suggests that an element of expertise is the ability to exploit externally-available information, thereby reducing load on internal memory. The notion of display-based performance is discussed further in Chapter 4, Section 4.2 where we discuss the relationship between the information available on the interface display and the actions that users carry out. We also discuss Larkin's theory in greater detail in Chapter 8, when we examine the extent to which user skills are based on the exploration of interfaces.

Many interfaces offer 'parallel dialogue styles', in that the users can select from a menu using a pointer device or issue an equivalent command from the keyboard. The advantage for frequent and experienced users of using keyboard commands to carry out menu selections is that they are generally faster, and do not force the user to switch between keyboard and pointer input devices during task performance. Where software is designed according to a common 'house style' determined by computer manufacturers (e.g., Apple's Interface Design guide, 1987; IBM's Common User Access guidelines, 1989), learning keyboard commands can be advantageous even for infrequent users. For example, almost all Apple Macintosh applications use <command S> to save the current work, <command X>, <command C>, and <command V>, to cut, copy and paste text or graphics, <command Q> to quit the application, and so on (the <> symbols refer to keys pressed at the same time). Thus, even when an application is new and the menus are unfamiliar, the chances are that one can apply one's knowledge of keyboard commands (though this, of course, relies on software developers following the style guides accurately).

As with question-and-answer and form-filling dialogue styles, there is a general belief that menu systems are preferred by inexperienced users and command languages by experts. There seems little strong evidence to support this. Indeed, Whiteside, Jones, Levy and Wixon (1985) found precisely the reverse relationship, with experts preferring menu systems, although we should beware of generalizations from the comparisons of particular interfaces. It is clear that we *can* create circumstances under which experts would find the use of menus annoying and slow. In these circumstances, it is interesting to investigate the circumstances under which users prefer to learn the keyed commands compared with selecting their menu equivalents. We could, for example, compare interfaces in which a function in the first is achieved with a single short command and in the other with several menu inputs. No prizes would be given for guessing the user's preference. It is to avoid this that products such as *macros* have appeared on the market, which overlay menu systems with 'speed-up' operations which have the effect of allowing several menu choices in a single keystroke. The provision of systems that the user can tailor to suit their own needs provides a potentially powerful solution to the requirement for context-sensitivity in dialogue design. However, they also impose the complex task of interface programming upon users.

2.6   Hybrid dialogues

The use of graphics-based menu systems has been made effective by the introduction of the mouse as an input device. The mouse is a pointing and selection device which has made available the possibility of selecting and moving objects around a display. This has revolutionized the kinds of question-and-answer, form-filling and menu systems that can be used, and has led to the widespread use of WIMP interfaces. It has also introduced a form of interaction known as *direct manipulation*. We discuss direct manipulation in greater detail in the following chapters, particularly Chapters 5 and 8, where we consider the means whereby users maintain control of dialogues.

The first major commercial system employing direct manipulation was the Xerox STAR interface, described by Smith, Irby, Kimball, Verplank and Harslam (1982). In this interface the act of selecting an item, usually represented as an icon, makes it the object of subsequent operations. Furthermore, items can be transferred from one location to another to denote printing (move to the printer), copying (move to another storage device) or deletion (move to a 'wastebasket'). Such a dialogue can be construed as hybrid because it involves many dialogue styles, such as the use of pull-down menus, forms, and responding to instructions or questions. In practice, direct manipulation cannot be seen as a distinct style of dialogue – it is difficult to imagine how a 'pure' direct manipulation dialogue would take place without using other dialogue styles.

Supporting task components with hybrid dialogues

As we have already remarked, most interfaces embody a mixture of dialogue styles. Indeed, it is hard to imagine an interface which reflects a 'pure' dialogue style. Even systems like PRESTEL allow for methods of direct access in which users can specify an information page directly without going through intermediate menus; cashpoint interfaces often involve the use of menus or the selection of choices from a range of function keys; and a typical microwave cooker controller necessarily combines menu systems for pre-determined cooking cycles as well as an overriding programmable sequence to allow for all other contingencies. This is directly analogous to the way in which human-human dialogues change style according to the purpose of the conversation.

As another example, Figure 2.9 presents part of a human-computer dialogue for paginating documents. This shows how form and menu dialogue styles can become combined. It also illustrates the sensitivity of the interface to the different types of input required. Aspects of the task requiring numeric input are pre-filled with default values which the user can change with keyed input; binary choices, such as Text Smoothing are selected by a mouse-activated 'on/off' button (crosses indicate activation); selection of a range of discrete possibilities, such as paper size, is enabled via a pull-down menu, currently showing the pre-selected default 'Tabloid'.

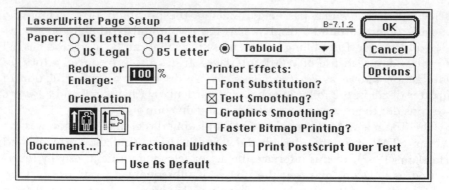

Figure 2.9 The combination of form and menu elements (where menu selections are made by clicking on 'radio' buttons) in a Print dialogue of Microsoft Word.

2.7 Conclusions

In this chapter we have examined how an interface can be designed to structure the tasks faced by users and remove some of the burden of planning and remembering task sequences and commands. We have seen how there is a complex relationship between the task that a user faces and the dialogue style that is most appropriate to support it. To some extent the choice of dialogue style will be influenced by the level of user skill and frequency of interface use. However, the primary decision about which dialogue style to choose should be based upon the nature of information flows within the user's task. We return to the question of how to analyse tasks in Chapter 9.

A question-and-answer dialogue is most appropriate for tasks in which the user must supply answers containing discrete bits of information in response to sequences of prompts. A form-filling dialogue is most appropriate for tasks which require the user to supply sets of related information, and in situations where maintaining the natural integrity of a body of information is important. A menu-based dialogue is most appropriate when the task requires the user to choose between system-prompted sequences of choices, such as in information-retrieval and command-issuing tasks. Whilst it is popularly assumed that these dialogue styles are suited to novice users and are less appropriate for experts, we have argued that the suitability of each style is context-sensitive, and is determined by more than just the level of user expertise.

As a final point, it is important to remember the interdependence that exists between dialogue styles and the available interface technologies. This is particularly evident in hybrid dialogue styles. The quality of display, the use of the mouse, and the available computing power (often tailored towards specialized screen-based performance) all enable an increasingly sophisticated range of devices to be used. In the next chapter, we examine how the *medium* through which human-computer interaction takes place exerts a significant influence on the nature of dialogues.

CHAPTER 3

Fitting dialogues to the medium

Overview

This chapter discusses the relationship between interface dialogues and the physical media through which they are conducted. The first section concentrates upon screen displays as a source of information to the user. The bulk of this section concerns the psychological aspects of system design which relate aspects such as formatting, colour coding, icons, clarity of screen display, and layout of information to theoretical issues pertaining to psychophysics, perception and the psychology of information processing. These theoretical issues reflect the cognitive concepts introduced in Chapter 1, Section 1.3. We draw a distinction between these psychological phenomena and the skills that users develop to exploit the information content of screens in complex tasks, which are discussed in Chapters 6–8.

The focus of the first half of this chapter is on the properties of screen displays, in which the flow of information can be characterized as passing from the computer to the user. The second half concerns itself with information flow in the opposite direction – from the user to the computer. The design of input devices offers different opportunities for, and constraints upon, human-computer dialogue. Some input devices have *enabled* new dialogue styles and even led to new ways of doing things, whilst other input devices, by their very nature, constrain some tasks to specific dialogue styles. The second half of this chapter illustrates one of the central themes in this book – that there is an interdependency between the user's task, the dialogue style designed to support that task, and the interface technologies which are the material link between users and system.

3.1   Display design

We devote a large section to the topic of display design because displays are, and will continue to be for the foreseeable future, the principal medium through which machines communicate information to their users. There are three main features which contribute to the design of screens. The first of these is the screen *layout*. Consider the two screen formats shown in Figure 3.1. The first screen is generally seen as poorer than the second. What makes the second screen better, and could these benefits have been predicted in the original design? What effects do good or bad layouts have upon the users? To answer these questions we need to understand the way in which the human visual processing system uses the layout of the display to guide the search for the required information.

```
ACCOUNTNUMBERS
431056,648792,134523,96734,342,5476
ADDRESSES
3,THE ROOKERIES,CHARLESTON ST,BIRMINGHAM.16 ROGER ROAD
HEMEL HEMPSTEAD.12 SINDERBY STREET LOUGHBOROUGH.234 ASHBY ROAD
WHICKAM.DUN ROAMIN, ENFIELD STREET, NORTHWOOD.10 DOWNING
ST SW1.
CHARGING RATES  7,7,14,70,7,40
DATE OF ORDER      23/10/91,12/12/91,14/11/90,20/1/91,9/9/91,
11/10/91
```

Account number	Address	Date of order	Charge rate
431,056	3, The Rookeries, Charleston St, Birmingham	23/10/91	7
648,792	16, Roger Road, HemelHemptstead	12/12/91	7
134,523	12, Sinderby St, Loughborough	14/11/90	14
96,734	234 Ashby Road, Whickham Dun Roamin, Enfield Street Northwood	20/ 1/91	70
342	10 Downing Street London SW1	9/ 9/91	7
5,476		11/10/91	40

Figure 3.1 Examples of poor and good information layout.

The second feature which contributes to the design of screen displays is *representation*. Suppose we wish to represent the yield of crops per square kilometre in a prairie. We could represent these data in tabular form, or graphically with the visual density of each cell representing the yield, as we do in Figure 3.2. How do these different forms of representation affect the way people think about the data? Clearly the way information is represented on displays has consequences for subjects' performance. The designer of screens might want to consider whether the primary aim of the interface is to support rapid scanning for configural information (e.g., there appears to be

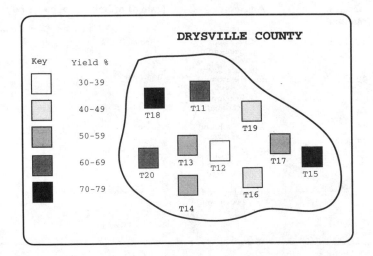

SITE	YIELD %
Site T11	63
Site T12	34
Site T13	55
Site T14	51
Site T15	76
Site T16	44
Site T17	57
Site T18	71
Site T19	47
Site T20	68

Figure 3.2   A simple comparison between tabular and graphic representations.

a 'dead' area of yield in Figure 3.2) or to enable exact readings. Different tasks seem to require different forms of representation. For example, Benbasat, Dexter and Todd (1986) found that decision-making tasks require graphical or tabular forms of representation depending upon whether the purpose of the display is to support the analysis of general trends or for explicit computation. The design of screens should reflect the purpose for which the user wants the information, and it should support the psychological processes that are undertaken to achieve that purpose.

The third feature of screen displays is the *coding* of information. Coding is a process in which a physical dimension (such as colour) is used to denote, often arbitrarily, some meaning to another object with which the code is associated. For example, Figure 3.3 shows how numbers, size, sequence, or a combination of these codes, can be used to denote the priority with which files have been accessed. Colour-coding is particularly common in everyday life (traffic lights, petrol pumps, taps, and subway maps all use this), but other dimensions are also available to the screen designer, including size, position, intensity, typography and graphics.

3.2　The layout of information on displays

The layout of information determines the ease with which the human visual system can navigate around a display. To understand how the layout of displays relates to the psychological processes of the visual system, notably those concerning the role of eye-movements, let us examine the task of reading text from a screen.

Reading text and the control of eye movements

Displaying continuous text on a screen would seem a straightforward matter. However, it is commonly recognized that reading from computer screens is slower than reading hard copy. This may be accompanied by fatigue or visual discomfort, inaccuracy, and lower comprehension rates.

These effects could arise from a variety of possible causes, such as the specific effects of posture or lighting which apply when reading from screens. Factors associated with the environment in which the task takes place are discussed fully in standard textbooks on ergonomics and workstation design (e.g., Salvendy, 1987). Dillon, McKnight and Richardson (1988) review a wide range of ergonomic and psychological factors, all of which they argue may contribute to the observed disadvantages of reading from screens. These include orientation of screens, eye movements, visual angle of view, the aspect ratio of the screen, flicker, polarity of screens (black-on-white or vice

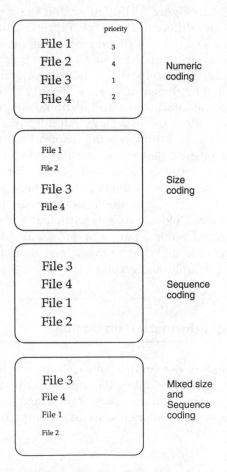

Figure 3.3   Examples of different coding dimensions for indicating the priority of file access.

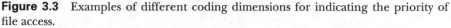

versa) and anti-aliasing, amongst others. Doubtless these factors are important. However, we would argue that many of the factors considered by Dillon *et al.* (1988) are general to all interface tasks, and do not pinpoint the special reasons why reading from screens is hard.

Recent work has focused upon the control of eye movements in reading text on screens, and suggests that it is the disruption of eye movements which leads to slower reading. The control of eye movements offers a useful, and general, framework for understanding the issues involved in designing screen displays. Eye movements in reading do not occur, as we might expect from the nature of written material, in a continuous flow from left to right over the text. In fact, the eyes move in a succession of short rapid movements (called *saccades*) followed by resting periods, or *fixations*, such that the eye jumps from fixation point to fixation point. Since the fixations last on average about 250 milliseconds, while saccades only take about 30 milliseconds, it is the *number* of fixations per unit of text that is the major factor in reading speed. This appears to vary according to both reading skill and the density of the text. Skilled readers jump larger distances in the text and therefore require fewer fixations per line of text than poorer readers. Reading speeds also differ between easy reading, such as a light novelette and more complex reading, such as an academic textbook. An easy piece of text containing short sentences and familiar words of only a few syllables, is generally read more quickly because fewer fixations are made, and they can be more widely spaced in the text. This presumably reflects the low information content and large amount of essentially redundant information of the text, whereby the gist can be extracted from fewer fixations. Complex material, on the other hand, contains much less redundant material and therefore requires more careful examination.

Research by Kennedy and Murray (1991) has shown a direct link between reading at CRTs (Cathode Ray Tubes, the mechanism by which images are projected onto screens) and the control of eye movements. CRTs are a display technology which is inherently prone to flicker. They show that these saccades are disrupted by flicker, even at 100Hz, which is considerably above the rate beyond which flicker is thought to be discriminable. The effect of this flicker is that saccades must be shorter, requiring more movements per line of text, hence reducing reading speed. Wilkins (1984) has also suggested that flicker may cause visual discomfort, due to the sensitivity of some users to stimulation at certain frequencies.

Peripheral visual information and eye movements

Control of eye movements is required in scanning all forms of visual information, be it graphics or text. The problem in this control is that the area of our visual field within which detailed information, such as text, can be discriminated is relatively small. If a stimulus is more than a small angle outside of the centre of fixation, our ability to decode fine detail diminishes. Yet the eye movements we make are often to targets *beyond* this range. How is this done? The answer is illustrated in Figure 3.4. While our peripheral vision is not as detailed as it is at the fixation point, it appears that broader features, such as the length of a word or word boundaries, are discriminable, and it appears that this crude information can be the target for the eye movement.

Reading is a specialized form of visual processing because it requires serial scanning of the written text, such that eye movements are generally all in the same direction. Most of these movements are relatively short because the next chunk of information is adjacent to the previous fixation. Other types of visual information are not intended to be viewed in quite the same way, and consequently, the eye movements are less systematic and may require the

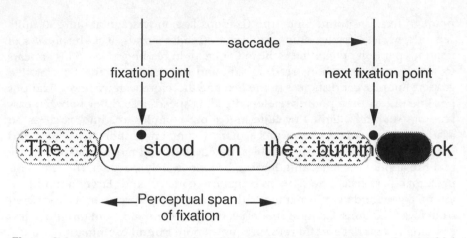

Figure 3.4 In reading, the legibility of text is preserved for a perceptual span which extends further to the right than to the left of the fixation point. Beyond this, legibility diminishes until only broad features are detectable.

reader to make substantial jumps across the display. Consider the second layout in Figure 3.1. It represents sales data for a number of clients. Comparisons of data across the display can involve large eye movements. How are they controlled?

A study by Lansdale, Jones and Jones (1989) illustrates a difference between reading text and scanning visual displays. They compared search times for locating information on displays which was represented in one of two forms: either pictorially or verbally. For example, a target could appear as the words "LEGAL MAIL" or it could appear with a pictogram of a letter, (representing mail) inside a circle (representing Legal). Care was taken to ensure that the pictures and labels occupied exactly the same areas on the screen per item. Subjects were carefully trained to recognize the different labels and pictures to ensure that the experiment measured speed of visual scan rather than speed of recognition. Lansdale *et al.* (1989) found that search times were reliably faster for the pictorial representations than for the words. Analysis of the subjects' search patterns suggested that, while subjects scanned verbal items serially, they were able to detect *pictorial* targets in their peripheral vision and so move towards them directly. A higher rate of missed targets in the pictorial condition is also consistent with the view that subjects were not scanning the pictorial stimuli serially.

The difference between scanning words and pictograms leads us to ask how these representations are significant for the design of displays. A major difference between the visual and verbal stimuli in Lansdale *et al.*'s experiment lies in their detail. Words of the same font represent a stimulus of high but consistent *spatial frequency,* that is, they present a fairly uniform array of lines and spaces between lines. A picture presents a rather more variable stimulus, with some areas having high spatial frequency and other areas having lower spatial frequency. A simple illustration of high and low spatial frequencies is illustrated in Figure 3.5. Imagine drawing a line across either of these displays and marking where the superimposed line intersects a boundary of an object (be it a letter edge or a part of a drawing). With text, the marks would be frequent and closely packed; hence the high spatial frequency. With pictures, some of the marks would be close, but others would be more spread out; hence the variable spatial frequency.

To explain the results of their experiment, Lansdale *et al.* argue that, whereas few of the verbal labels were discriminable in peripheral vision, many of the pictures were identifiable by the variable spatial frequencies which were detectable in peripheral vision. This argument is based upon the fact that, as targets move from the centre of our visual field (defined as the fixation point) to the peripheral areas, it becomes harder to discriminate details. At the edge of our visual field, we are sensitive only to low spatial frequencies which represent coarse levels of detail. Therefore, subjects scanning in the pictorial condition had more information as to the location of certain targets than did subjects in the verbal condition.

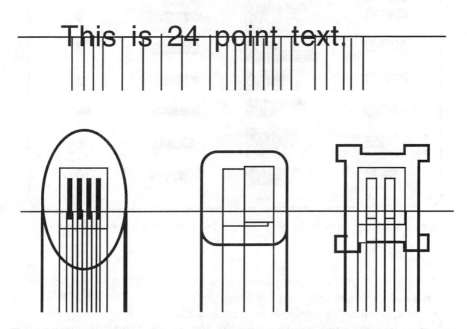

Figure 3.5 A spatial frequency analysis of text and pictograms. Note how the text produces a relatively consistent array of spatial frequencies, whereas the pictograms are more diverse, with some large gaps between elements as well as some extremely small ones.

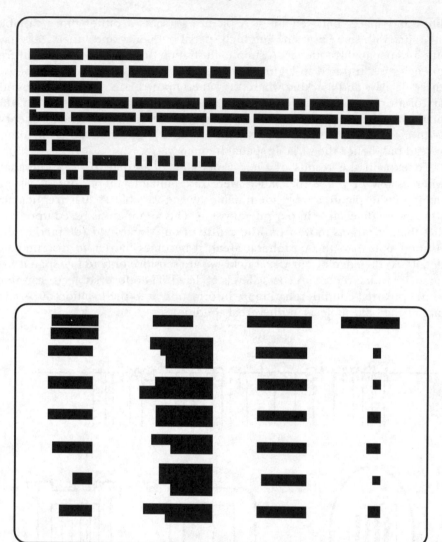

Figure 3.6 A degraded version of Figure 3.1 showing only low spatial frequencies.

The suggestion that we use spatial frequency information in our peripheral vision to assist visual scan has been made before (e.g., Findlay, 1981), and suggests a number of methods of screen design which can help the user. Consider again the two displays shown in Figure 3.1. Why is it easier to find data on the second version? Figure 3.6 shows a blurred vision of these screens which highlights their low spatial frequencies. The first version presents relatively featureless blocks. The second version, by dint of the blank columns and rows, preserves the basic structure of the display and presents clear targets which the visual system can use in moving to the required point.

A basic principle in screen design is therefore that screens will be easier to scan if the layout allows for some discrimination in peripheral vision. This explains the findings that simple icons are more quickly scanned than complex ones (Byrne, 1993). The same applies to many of the factors influencing reading speed that were identified by Dillon *et al.* (1988). For example, 'font resolutions' (basically, how jagged the characters are) are directly related to the spatial complexity of peripheral vision. The clearer the font resolution is, the more readable the text will be.

Many design guidelines also reflect the principle of discrimination in peripheral vision. For example, it is often recommended not to right-justify text. We can now see that right justification has two consequences: it removes broad visual cues from the ragged right-hand edge of the text; also it can create uneven spacing between words which may interfere with normal reading eye movements. Newspapers solve this problem by keeping their columns narrow. Another example is the need for spacing between text paragraphs, rather than running them from the line following the previous paragraph. This has the effect of introducing discernible peripheral cues which enable the reader to jump to the beginning of paragraphs easily. (Note that the aesthetics of publishing sometimes override these guidelines.)

Layout and perceptual grouping

Thus far we have considered the processes which control eye movements. But what leads the user to look at one thing rather than another? In this section, we examine how users come to conclusions as to the structure of information on the basis of its layout. Consider again the second display in Figure 3.1. It is evident that the items in each column are connected in some way and are differentiated from the other columns. Within each column, the contents are also clearer. For example, in the first display the addresses are hard to separate, whereas in the second display they are clearly separable. There are two ways of inferring this from the display. We can simply look at the items to discover from their contents that they are related. That conclusion can be reached also from the visual layout of the page. However, difficulties can arise when these differing sources of information conflict.

Why do we infer content from appearances? German psychologists of the Gestalt school (e.g., Wertheimer, 1958) showed earlier in the century how our perceptions are influenced by the structure that we place upon visual information. For example, Figure 3.7 is more readily interpreted as showing three circles over which a white triangle has been placed than three circles with missing segments. Similarly, subtle changes in relative density of dots are perceived as a boundary in Figure 3.8. In short, we appear to infer simple

Figure 3.7   An example of perceptual grouping. We see this as three black circles superimposed by a white triangle rather than as three circles with missing segments.

structures in the visual world wherever they can be found, a continuous process that takes place unconsciously and uncontrolled. This makes good evolutionary sense, since regularities and disruptions in visual patterns are most often correlated with objects or boundaries of some kind.

A rule of thumb is that if it is possible to infer an apparent structure from the layout of information on the screen, then the user will undoubtedly perceive it. The designer's job is therefore to ensure that the structure that is

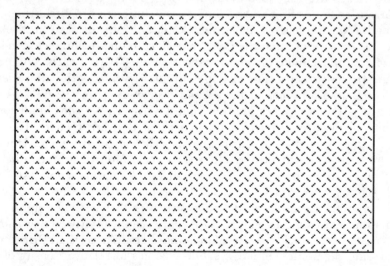

Figure 3.8   Small changes in texture are perceived as boundaries.

inferred is the one intended, and that misconceptions are not encouraged by spurious clues in the layout. For example, Figure 3.9 shows a help screen from a graphics controlling interface. It lists a number of available commands, and breaks just about every structuring rule one might think of. First, the relative spacing of the columns and rows leads to the perception that the information within columns is related – which it is not – and makes the reading of rows rather difficult. Second, the commands are listed partly alphabetically and partly according to function; but these groupings are in rows rather than columns. A consequence is that no search strategy can be relied upon wholly. Furthermore, the most appropriate strategy, of scanning along rows, is actually inhibited.

A second example of misleading layouts arises from the study of a menu-based viewdata system by Young and Hull (1982), which we discussed in Chapter 2, Section 2.5. A major issue in the design of these systems is the problem of users making incorrect choices from the range of alternatives offered, resulting in their getting 'lost' in the system. Young and Hull show how the chances of the user missing relevant choices is increased by the presence of misleading menu formats. For example, in one screen concerning train timetables, a number of options were laid out as a list (thereby strongly implying a set of mutually exclusive items) with further options embedded in accompanying text. Many users appeared to behave as if they had assumed that the list represented the full range of options and accordingly missed other options. In other words, subjects were assuming from the layout of the display that the list represented the menu options and the accompanying text was supplementary. This study is a good example of how, rather like leading horses to water, you can place information on the screen, but you cannot always make the user read it in the way intended.

AllocP	Arc	BFill	BFilS
CopyS	CopyTS	FFill	FFilP
IBCol	ICCol	ICP	ICDP
IFCol	IPat	IPCol	IRSel
ITCol	IWProt	Limage	LimagC
LineTo	LSym	LSymC	MoveR
Pinit	Plot	PlotR	PlotRS
RimagC	RPix	RpixR	RPixRS
SBCol	SCCol	SCDP	SCSP
SGFPatR	SFPats	SHires	SLores
SRscl	SStyle	STCol	SWProt
IDBank	Bcopy	SDBank	BRLut

Figure 3.9 A section from a graphics help screen showing a misleading columnar structure.

3.3   The representation of information on the screen

This section considers the effects that different representations have upon the user. A simple example is the use of either verbal or pictorial labels to represent functions or actions, as illustrated in Figure 3.10. Part of the difference between these representations is due to their differing perceptual clarity, a factor influenced by the 'preconscious' cognitive processing mechanisms of perception and attention. There are also effects of prior knowledge on the way these representations are interpreted. As the work of Young and Hull (1982) shows, there may be an interaction between preconscious processes such as perceptual grouping and the conscious assumptions users make about the information *content* of the screen. In other words, what the users assume about the information they are looking at is dependent on the way in which it is laid out. For example, a list of items in a menu dialogue carries the implication that it contains a complete set of possible options. For some people, this will be accompanied by the assumption that the items represent a mutually exclusive set of concepts.

Since different representational formats can lead to different interpretations of a set of information, it is important to check that an appropriate representation is chosen for a particular task. The work of Benbasat *et al.* (1986) discussed in Section 3.1 is a good example of this: judgements involving quantitative comparisons require tables whereas assessment of trends can be better achieved by graphical means. Another example is the organization of dial displays, illustrated in Figure 3.11. The purpose of the display is to signal abnormal readings to the monitoring controller. The dials are organized in such a way as to allow perceptual grouping to emphasize the abnormal reading by arranging the normal levels (regardless of their absolute values) to be consistent.

Representational formats can also be misleading, either in the information they withhold, or in the assumptions they falsely encourage. For example, Figure 3.12 shows two versions of a histogram which illustrate a mythical relationship between fruit preference and incidence of headaches. The first version has had its vertical axis adjusted to suggest that one fruit preference occurs with a significantly larger incidence of headaches than the others. By changing the scale, the second version emphasizes how alike the values are.

A rather more serious example is possibly to be found in the official report following the airline accident at Leicestershire, UK in 1989 (Aircraft Accident Report 4/90, Department of Transport, 1990). The engine vibration indicators were such that abnormally high readings were perceptually grouped with neighbouring dials which showed normal readings. Although not directly causing the crash (disasters are rarely attributable to a single cause), the crew's decision which resulted in shutting down a fully functional engine and relying upon a seriously damaged one was not helped by this arrangement of flightdeck indicators.

```
>LIST CONTENTS

RTHIST.DOC
RTHIST.SNO
RTHIST.DAT
PETER.SNO
TBGG.RES
>
```

Figure 3.10 Textual and pictographic representations of data files. Note how the pictograms aim to give feedback about the contents of the file. In Chapter 5 we use this as an example of how visual and location cues can increase the distinctiveness of objects to minimize the likelihood of incorrect selection.

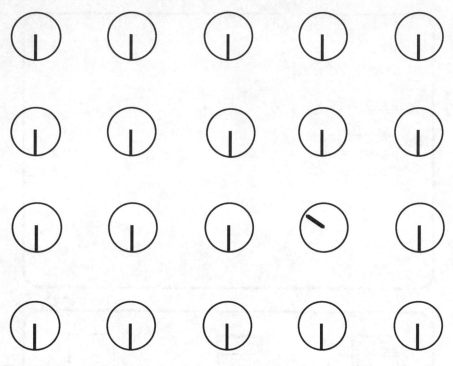

Figure 3.11   How perceptual grouping can be exploited to show abnormal readings clearly.

3.4   The coding of information

Coding is a process in which a physical dimension, such as colour, is used to represent a quality, quantity or change in another dimension, such as temperature. One issue is to consider is the ease with which the user can work out what the codes mean. Like traffic lights, the association between the codes and the intended information must be learned. How people learn these codes over time, and the difficulties they encounter in doing so, involve a complex interaction between their understanding of the task they are trying to carry out, their experience of the task, and the prior associations they have to the items. Most interfaces use implicit codes somewhere or other, such as the use of increased brightness to indicate a selected item, and one cannot rely on prior task experience or training to prepare users for all eventualities. Users must therefore be able to interpret codes as they encounter them. This leads to two questions: first, how many different values of a dimension can a user keep track of; and second, what determines how easy they are to tell apart (i.e., their perceptual discriminability)?

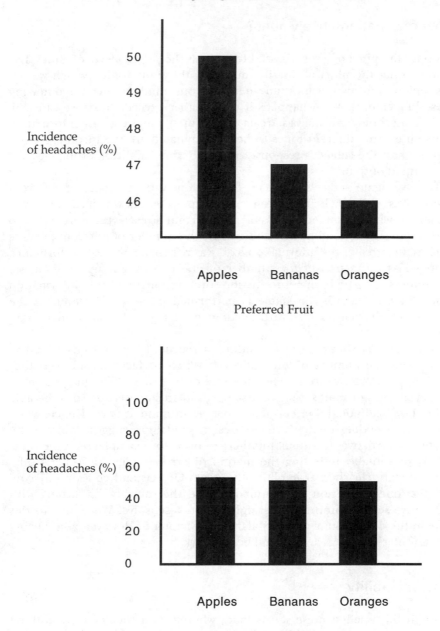

Figure 3.12 The misleading effects of different axes in diagrams, showing how a false impression of difference (or similarity) can be generated by a suitable choice of scale.

How many alternative values?

Suppose the pitch of a tone is used to express the priority level of a message. Disregarding the physical discriminibility of those intensities, which we discuss below, how many values can we hold in our minds at once when making these judgements? What happens if the number exceeds those we can handle? A psychological analysis treats the human subject as an information transmitter: the different tones to be discriminated are the incoming information; and the subject's responses to those tones, right or wrong, are the transmitted output.

To investigate subjects' capacity for dealing with numbers of different alternatives, Pollack (1952) conducted an experiment in which subjects were presented with a number of possible pitches and at each trial were asked to indicate those they had heard. The greater the number of different possible pitches, the greater the information load was per trial. Subjects' performance changed as a function of the number of alternatives. For up to about six alternatives per trial, the information output mirrored the information input. In other words, the subjects performed perfectly. However, as the number of alternatives rose further, then the subjects began to make mistakes.

Miller (1956) shows how this finding is repeated for all kinds of judgement, with the number of alternatives at which perfect transmission stops being approximately seven, hence the title of his paper: "The magical number seven, plus or minus two". In essence, human subjects appear to be able to hold only a limited number of alternatives in mind at once. Hence, when considering coding methods, many design guidelines suggest limiting the range of alternatives to a small number – 4 or 5 (for example, see Cole *et al.*, 1985), presumably reflecting the margin of error implicit in Miller's title. This estimate clearly oversimplifies the issue. Objects and codes rarely contain just one dimension of information, and the capacity for dealing with alternative sets of information is highly context-sensitive. We return to this issue in the discussion of auditory alarms in Chapter 5. However, as a starting point, Miller's findings are a useful benchmark.

Discriminability

At night the smallest noise seems loud, whereas at a rock concert nothing short of an explosion could increase the perceived overall loudness. What this tells us is that it is not the *absolute* volume of a sound which determines whether we notice it, but the *relative* increase over the background noise which matters. The same basic principle applies to other senses. For example, in sunlight additional lights need to be very bright to be noticeable, whereas in a dark room a flickering candle is seen as relatively bright. This property of our perceptual mechanisms to base discriminations upon relative judgements has clear implications for discriminability of codes at interfaces.

As a broad approximation, the additional stimulus required to make a signal detectable against a background source can be determined by Weber's Law (after Weber, who first stated the principle in 1834 – a detailed discussion can be found in Laming, 1986). It states that, for any stimulus to be discriminated against its background level, its intensity must exceed that of the background by a given fraction of the background intensity. This is given by the function:

$$\text{Discriminable new value} = \text{Old value} + k(\text{Old value})$$

where k is a constant. This means that when making a coding dimension discriminable from background levels, the necessary increase depends upon the context. Auditory alarms are a good example: if the background noise is low, as in a quiet control room, then modest volumes of sound will be discriminable. In a noisy environment, such as a factory floor, the same tone would be inaudible, and greater volume is required. Note also that Weber's Law governs differences which are on the threshold of being just noticeable: to be easily discriminable, additional intensities would need to be greater still.

There are a number of interacting factors which point towards using only a few levels of any single dimension (such as brightness or sound volume) used for coding. First, the different levels must be sufficiently spaced along the continuum of values so as to be discriminable from each other. Second, many coding dimensions, such as brightness, are limited by technology to a small range of values. For example, there is a limited range of luminance which CRT phosphors can deliver. Since increasing levels of illumination require increasing separation for discriminibility, this will rapidly limit the number of possible alternatives that are available if brightness is used as a dimension. Finally, when we add to these considerations the need for human comfort and safety, and the ability of humans to remember only a small number of these distinctions accurately, it is evident that coding methods in which users are expected to associate one dimension to another are limited to a small number of alternatives. This is reflected in many guidelines on coding, which emphasize the need to limit the number of alternative values per coding dimension (e.g, Williges and Williges, 1984). Doubtless users can be trained to make finer discriminations, just as musicians can learn to discriminate fine gradations of pitch, but violating this basic principle is always likely to invite confusion.

We have discussed the discriminibility of coding dimensions in terms of the ability to discriminate across single simple dimensions such as sound volume, brightness or size. However, these can be combined, as in Patterson's (1990) use of volume, timbre, rhythm and other sound attributes in the design of auditory alarms, which we introduce in Chapter 5 when discussing coded alarms as a form of feedback. The situation in this case becomes much more complex, and relies more upon the user's long-term expertise in the tasks for which such coding dimensions are used, such as medical monitoring systems or flying helicopters. Nonetheless, our description of concepts of short-term discriminibility and memorability are generally applicable to all coding methods, and provide the basis for considering more complex designs, such as auditory alarm systems.

Colour coding

Colour coding is an interesting area to focus upon because it touches upon so many psychological, as well as technological, issues. Reviews of its use in interface design are given by Travis (1991) and Shneiderman (1987). Colour coding is used to delineate areas on the screen, to attract attention, to denote the different status of items, as well as being used purely decoratively. On this latter issue, that of relating aesthetics to design, cognitive psychology has little to say, but on other issues we can do a little better.

The ability of colour to attract attention is a particularly strong, and is rooted in several psychological mechanisms. First, given the low resolution of peripheral vision, blocks of colour are one of the features (provided they are not too eccentric to the line of vision) whose appearance can be detected in peripheral vision. Given our discussion in Section 3.2, colour coding of blocks of tables, for example, can considerably speed up visual search.

In some circumstances, colour can be sufficiently attention-grabbing as to be intrusive, and appears to act preconsciously upon users. An example of this is the 'Stroop' phenomenon (Stroop, 1935). This is an effect of colour recognition intruding upon another cognitive process. It is demonstrated experimentally by asking subjects to search lists of names of colours to select instances of a particular name, such as RED. If the names are printed in conflicting ink colours (such that the word RED is written in blue ink), then speed of searching is greatly increased and subjects are more likely to select words incorrectly which are written in the target colour. The colour of the target appears to conflict with the name, thereby disrupting the subject's conscious processing of the colour names.

Some colours are strongly conditioned by prior experience to have strong associations, such as *Red* for stop or danger, and *Green* for safe or go (e.g., see Travis, 1991). The implication for interface design of the Stroop phenomenon is that these associations can impose themselves upon our attention. If these conflict with the interface functionality, such as, intended using *green* for danger, then errors may follow. These colour associations are sometimes very subtle. For example, some people implicitly code colours as being warm or cold, and may confuse colours within those categories. Coloured objects upon screens may equally be classified implicitly as being in opposing or related sets, especially when colours are presented that are implicitly related (such as orange and yellow) as opposed to two which are antagonistic, such as red and green, or red and blue. Violation of such prior associations can cause difficulties because the long-term associations that people develop for certain colours will constantly interfere with the task at hand. The Stroop phenomenon shows how colour literally 'captures' peoples' attention. It follows that if the user of an interface misinterprets the colour coding according to prior expectations, then errors can occur. At the very least, colour coding can be a distraction. Rigg and Sandringham (1977), for example, report subjects' frustration at inconsistent and irrelevant use of colour coding.

Colour can therefore be used as a means of differentiating areas of the screen; it can be attention-grabbing, and it can imply and reinforce subtle associations or distinctions between visible items. Conversely, if used inappropriately, it can be distracting and disrupt visual search when the implied structures do not correspond to those intended, and can imply associations which may not actually exist. For this reason, guidelines often warn designers to be sparing in their use of colour. Cahill and Carter (1976), for example, suggest that from the point of view of speed of visual search, any more than four colours causes confusion, whereas with three colours, a target colour is perceived by a human subject as 'standing out' from the alternatives.

There are other problems in the use of colour. The most obvious of these is that a substantial minority of the population (about 8% of men, though fewer women) are colour-blind, in particular failing to differentiate between red and green. Second, highly saturated colours (that is, 'pure' colours which have not been mixed with white to give a washed-out appearance) do not focus exactly in the same plane. Consequently, mixing saturated colours of different wavelengths (e.g., red and blue) is likely to make one of the colours appear out of focus, or to be on a separate plane (chromosteropsis), and may possibly be accompanied by visual discomfort.

A final consideration is that the phosphors used to produce colour on CRTs do not behave in exactly the same way, partly because of their hue and partly because their photochemical properties differ slightly. They vary both in terms of their perceived brightness and in their degree of flicker. The latter depends upon the speed of recovery of phosphors as the scanning electron beam moves over them. Slower phosphors, such as green, flicker less because their luminance has reduced less between successive scans than faster phosphors such as blue. To those sensitive to flicker, blue horizontal lines in peripheral vision (which is anyway more sensitive to flicker than central vision) are far more uncomfortable than the same lines in a slow green phosphor. Travis (1991) gives a good account of the range of difficulties that can occur through flicker. Thus the effectiveness of colour in interfaces is particularly sensitive to the particular technology used.

3.5   Summary: Display design

The process of visual search is based upon the ability to find reliable and discriminable indicators of structure in the display. These include the use of boundaries, colour coding, and partitions. Their shared property is that they are discriminable in peripheral vision. To be effective, the structure that they imply must be consistent with the actual structure of the data. Structure is implied in a number of ways. Users may have to learn the associations between codes and their meaning. Structure is also shown by related colours, perceptual grouping (which links objects by their locations), or by being contained in a common area. In all cases, users *actively seek meaning in structure*. If that structure is inconsistent, arbitrary, or cuts across prior associations, then errors may result.

Users' recognition of codes and structures is dictated by their ability to discriminate between alternative values and to hold the separate values in mind. Without training, it can be difficult to discriminate anything but a small numbers of alternatives. For this reason, some guidelines suggest that the use of mechanisms such as colour coding should be minimized wherever possible.

In this section we have focused upon visual displays since they are both better understood and easier to explain in the context of interfaces than other output media. Nevertheless, many of the general psychological issues we have discussed apply to auditory messages and other forms of output. Some of the recent research into auditory output is reviewed in Chapter 5 where it is discussed as a form of feedback.

3.6   Technological aspects of dialogues: Input devices

Descriptions of input devices tend to date very quickly, and in so doing, can fail to get across generalities which will apply to future, as well as current, methods. Here, therefore, we introduce a sample of input methods in order to give a flavour of the issues which apply when considering the merits of specific methods in particular circumstances. We do not discuss the use of scanners and optical character recognition for inputting text data, because their use is largely determined by technological constraints. Also, we do not discuss empirical evaluations of various input devices, although there is an enormous literature in which almost every conceivable configuration of devices and their operating parameters is evaluated. Whilst such evaluations provide useful information for the developer trying to weigh up whether to use one device or another, they are of little long term value because technological progress means that benchmarks are either overtaken or become irrelevant. Surveys of input devices are to be found in Shneiderman (1987) and Helander (1990), and reviews of these technologies often appear in popular journals.

Keyboards

A large number of ergonomic studies of optimum keyboards have been carried out, concerned with the angle of keyboards, pressure required to activate keys, layout, and so forth (see Salvendy, 1987, for a review). Keyboards have become a microcosm of ergonomic study in their own right, with a vast literature reporting the benefits of particular layouts (for a review of research on layouts see Potosnak, 1988). Whilst study of the ergonomics of keyboards has improved their quality, research into keyboard layout has had virtually no impact upon the design of commonly used interfaces. This is not because the research is poor quality, but because its outcomes are frankly irrelevant to the majority of keyboard users.

Keyboard layouts are a good example of how the constraints of early technology can lead to a design which is, technically speaking, sub-optimal. The QWERTY layout (referring to the sequence of keys on the left hand side of the top row of characters) was designed to *slow down* the typists by strategically spacing out common letters. This was necessary in mechanical typewriters because high typing speeds caused a jamming of the striking arms. This constraint no longer applying, several attempts have been made to redesign the keyboard. The simplest approach has been to rearrange the keys, such as in the DVORAK design. This might also be accompanied by an ergonomically more desirable keyboard shape, such as in the MALTRON design. Chord keyboards, in which simultaneous pressing of several keys can represent word parts or whole words represent another design alternative.

There is no doubt that trained users (such as stenographers) can increase their performance by orders of magnitude with these devices (Shneiderman, 1987). However, these improvements have, so far, had little commercial impact. The vast majority of computer systems have retained the QWERTY keyboard. One reason for this, other than simple inertia in the market, is that the QWERTY keyboard is *good enough* for most users, and that the advantages of other keyboards have not been great enough to outweigh the following factors:

i. *Retraining and capital costs.* Virtually every computer and typewriter in every office, school or home has a QWERTY keyboard. Add to this the fact that everyone who has used a keyboard has learned the QWERTY pattern, the economic costs of introducing a different keyboard are prohibitive;

ii. *Marginal added value.* It is questionable whether many users would actually benefit from a change of keyboard because they are not, and never will be, skilled typists. Whilst it is possible for an ergonomist to produce objective evidence that keyboard X can allow faster typing speeds than keyboard Y, fast typing speeds are irrelevant to the needs of many computer users. In the future, it may be established that more ergonomically designed keyboards significantly reduce the incidence of serious conditions such as repetitive strain injury. If this is accompanied by appropriate legislation, then the added value for keyboards such as the MALTRON design might increase;

iii. *Optimum keyboards are specific to languages and tasks.* Optimum keyboard layouts are chosen from a consideration of the proximity of letters in text. In order to make such analyses, one must consider the frequency of particular letter pairings within the texts that users will type. However, this varies according to the nationality (for example, English has different letter pairing commonalties than French and German) and the nature of the material to be typed (medieval history and nuclear physics employ vocabularies which make vastly different demands on keyboard characters). If a specific task requires specific inputs (such as chemical or mathematical symbols), this may well have an impact on an optimal layout.

Mouse and tracker-ball inputs

For many years, the keyboard was the sole method for users communicating with machines. For this reason, dialogues evolved around typed responses from the user. The mouse and a host of other pointing devices (light pen, touch screens, joystick, tracker-ball) changed this because the user could select by location on the screen rather than by description.

The advantages of the mouse over other methods of input are illustrated by the following example: suppose, in a directory of files, one wants to select a subset of files to be copied to another disc. In a traditional keyboard system, this would require either typing out the filenames or descriptors thus:

copy c:file2 d:file2
copy c:file3 d:file3

In comparison, a single sweep of the mouse accompanied by a pressed button can serve the same function. Pointing is much easier than typing, and moving an object from one file location to another is easier by 'dragging' (selecting a screen-based object with the mouse and moving it by moving the mouse while the button is held down) than it is by typing in the appropriate command. Typed commands can require both source and destination to be typed in, rather than simply pointing to their location.

A mouse, in conjunction with high resolution displays upon which detailed images such as icons can be displayed, adds an entirely new range of user inputs to the interface. Much of this involves pointing at objects, selecting them, and moving them to other locations. This is the basis of direct manipulation. The mouse has consequently been central to a number of new interface design philosophies, notably the so-called WIMPS (Windows, Icons, Menus, Pointer Systems) interface style. An example is shown in Figure 3.13. The preoccupation with manipulating objects and and their locations is intentionally literal. It is usually coupled with the WYSIWYG approach to design: What-You-See-Is-What-You-Get; in which screen formats are deliberately kept as close as possible to the appearance of the same material when printed out.

Much has been written on the relative merits of different pointing devices (e.g., Shneiderman, 1987; Card, Moran and Newell, 1983), but it appears that the mouse is as good a pointing device as any other, and certainly better than some (Card, English and Burr, 1978). It is interesting to note that a mouse is not in itself sufficient to provide an enhanced interface. As we noted in Chapter 2, WIMP interfaces are complex in terms of their dialogue styles. They incorporate direct manipulation, menu selection (facilitated

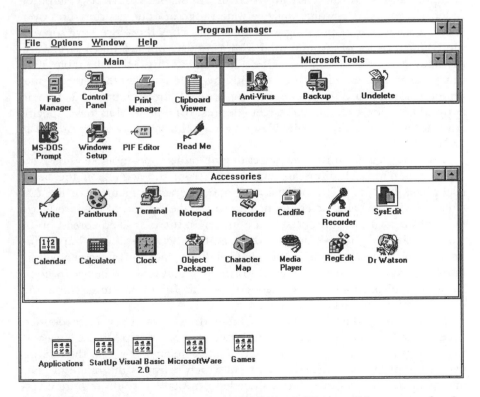

Figure 3.13 The program manager from Microsoft Windows 3.1, an example of a WIMP (Windows, Icons, Menus, Pointer) interface.

because the menus are usually accessible without changing the current screen and can be selected using a mouse), question-and-answer, and form-filling dialogues. The value of the mouse for WIMP interfaces would seem to be twofold. First, they provide the gesturing facility which is an essential component of the WIMP approach to design. Second, dialogues such as menus and question-and answer can be more easily used because a user can select a choice directly rather than having to express it through key presses.

Tablets with dedicated commands

Tablets are flat platters, often considerably larger than the system screen, which can detect the location of a stylus or other device placed on their surface. Generally, the platter is laid out such that different areas denote different functions, and they are overlaid with graphics to indicate the functional areas. Design issues for tablets are rather like the design issues for screens, with the aim to ensure that the dimensions are easily scanned, consistently grouped, and so forth. They are used in applications, such as Computer-Aided Design (CAD), which utilize a wide range of options printed on overlays to be selected by pointing with the stylus. They therefore have some of the characteristics of a menu used with a mouse, but are much larger and permanently visible to the user. One advantage of this is that the tablet itself can act as an *aide memoire* to the user with the added benefit that the user may remember not only the name and purpose of the function, but also its approximate location on the tablet. Because of this, tablets have specific advantages for visually impaired users, where the tablet overlays can be in braille.

However, two negative characteristics of tablets are notable. First, the user's command selections are made away from the screen on the tablet itself. This has the effect of separating the interactive element of the task from the objects being acted upon. This can make control of the interface a more abstract, indirect process. It does not help the user keep abreast of the current status of the task because the feedback between what is happening on the tablet (itself minimal) and what is taking place on the screen is itself indirect. Second, to make the tablet worth having, it tends to be used when a large number of functions are encoded upon it. Tablets are most commonly seen, therefore, in niche markets such as CAD.

Graphic tablets have a number of other drawbacks. They are expensive, occupy a substantial amount of desk space and are not particularly easy to use. The question is why are they used at all? The answer is that the alternatives are even worse. To design a system which provides the range of commands or parameters that can be encoded on a tablet would require a very complex command language (tying the user to a keyboard without the option of the *aide memoire* which a tablet provides) or a pull-down menu system of labyrinth-like complexity which would increase the complexity of the dialogue in other ways. The use of a tablet illustrates how some tasks are difficult to provide easy-to-use interfaces for, and the solution is merely a compromise which minimizes the unsatisfactory attributes of other input methods.

Tablets for freehand input

An obvious development of tablets is to allow freehand input, and a number of such products are beginning to appear on the market (see BYTE magazine, February 1991). These offer a number of advantages, especially if freehand text recognition (a notoriously difficult thing to program computers to do) becomes reliable. First, the need for a keyboard diminishes. The input device, usually some form of stylus, could act both as a means of cursor control as well as textual input. This will be of particular significance to portable systems. Second, it removes the need to learn keyboard skills, which for many (but probably not all) circumstances will mean an increase in efficiency. Finally, it offers the potential to make many expressive gestures available as command inputs. For example, all the conventionally agreed proof-reading signs could make the editing of text much easier by freehand input than other methods. In future years, it will be interesting to observe the niche that freehand input comes to occupy in the range of tasks for which different input devices are used.

Touch screens

Touch screens are rather like vertical, screen-based graphics tablets: the user points to objects upon the screen either with their finger or with a pointing device like a stylus. There are a number of different technologies for touch screens, and these give rise to comparisons with touch and tablet methods, such as the observation that the hand obscures the screen, and vertical screens are less comfortable to use or easy to be precise with (Shneiderman, 1987).

Touch sensitive displays focus the dialogue onto the screen itself. Consequently, to initiate a touch-based interaction requires the giving-up of screen space to enable the user to see given targets. It follows that either a section of the screen is dedicated to touch input, or that the touch-input dialogue is modal (i.e., it is present at some times and not at others). This may be inconvenient if the task requires certain inputs to be possible at any time, or if a large number of inputs are possible. One solution is to use two screens. However, if a second screen is dedicated to touch input, one would want to ask whether a tablet would not be preferable anyway. If an interface requires a large number of possible options to be available simultaneously, then touch methods may not be up to the job. The standard way of cramming a large number of options into a small area is to make the touch areas reflect a hierarchical menu system. To this extent, the interface technology drives the interface design: touch screens might lead to hierarchical menus and tablets might avoid them.

Touching is a form of communication which users may find odd when there is little or no tactile feedback connected with the changes taking place on the screen. This is more acute with touch screens than with styluses, where mechanical mechanisms can be devised to give a clear indication as to when a press has been registered by the system. In touch screens, particular care has to be taken to give the user clear feedback as to the effect of their input (Potter, Weldon and Shneiderman, 1988). Even then, the impression seems to be that users find it disconcerting to match their pressing of a piece of glass with a change in state of the system.

A final limitation of touch screens and some tablets is the fact that the user can only make one kind of response: to press or not press on the screen (some tablet pointers are more complex than this because they can carry buttons). The 'bandwidth' of communication is also more limited than a mouse system because of the speed with which mouse button(s) can be pressed compared with the activation of the touch-screen, and because a mouse can have more than one button. This seems to limit the dialogue to a condition in which the user selects one of a range of choices and then make a secondary input to activate the selection. It follows that any activity which involves a number of contingent inputs, such as entering data on a form, is extremely cumbersome and requires a very large number of screen presses. Touch-driven dialogues are therefore most often seen for relatively simple tasks in which the user makes selections from a limited range of options.

3.7   Conclusions

The main theme of this chapter has been that the technology of input and output devices interacts with constraints on human information processing to determine the opportunities for designing human-computer dialogues. This principle can be seen to apply to output devices. Here we have concentrated upon the psychological issues relevant to the layout of information on visual displays. We have seen how the layout of information on a screen influences the way in which information is perceived, notably in terms of the influence of layout upon eye movements. Also, we have discussed how different representation can enhance or detract from the meaning behind information displays. We have also seen how limitations of the human visual system constrain the range and levels available for coding information into meaningful groups.

Input devices influence the technical opportunities for dialogue styles. Since dialogue styles are not independent of task, it follows that the link between interface input devices and tasks is not an arbitrary one. In some cases, tasks may require unusual or sophisticated input devices. However, the optimization of input devices can be limited by resistance to change on the part of users who may have invested heavily, in terms of money and training, in older technologies.

CHAPTER 4

The structure of dialogues

Overview

This chapter presents an analysis of the languages used in human-computer dialogues, and examines their relationship with the languages of human-human communication. First, we focus on the notion of *consistency* in dialogue design, an important dimension by which interfaces can be judged. We discuss ways in which consistency is maintained, both within interface languages, and between interfaces and users' existing knowledge. Task-action grammar analysis is adopted as a framework for assessing the consistency of both text-based and graphical dialogues, and we examine the extent to which this framework offers a useful method for designing and evaluating interfaces. The concept of *affordance* is introduced to describe consistency between the interface display and the user's understanding of an interface. We then discuss how the consistency of a command language's vocabulary and grammar affects its usability. Later in the chapter we explore a number of language-related dimensions other than consistency which capture properties of dialogues. Finally, the chapter examines the extent to which the natural language of human-human communication is appropriate for human-computer dialogues.

4.1   Introduction

In the same way that people communicate with each other using spoken or written language, dialogue with a machine also requires a language. However, human-computer dialogues are rarely performed as effortlessly as most dialogues between people, and usually require extensive training and practice. A study of the structures that underlie languages used for interfaces can uncover the potential difficulties in communicating with machines. Lately, the description of dialogues in terms of their linguistic properties has become less common, largely because of the increasing influence of graphical user interfaces, in which pictures and movement rather than words are used to issue commands. Nevertheless, dialogues without words can be described, by analogy, as having language-like properties. This chapter examines how far linguistic analysis offers a useful mechanism for describing dialogues for both textual and graphical interfaces.

Why might a linguistic analysis of dialogues be of value? As an analogy to learning human-computer dialogues, consider the problems faced in learning a second language. For example, it is generally recognized that Spanish is an easier second language than German for native English speakers to acquire. Similarities between the vocabularies of Spanish and English (deriving from common Latin origins) mean that one can quite readily guess the translation of simple Spanish phrases, by predicting the meaning of words based on their English equivalents (e.g., the meaning of "¿Quanto costa uno vino tinto, por favor?" should be fairly clear even to those who have learned no Spanish). The grammar of Spanish is less complex than that of English, in that it has a relatively simple set of rules for adapting verbs, nouns and other words to accommodate different tenses, subjects and so on. Unlike English, there are few irregularities that require special endings or rephrasings. Thus the relative simplicity of learning Spanish for native English speakers can be explained by its internally consistent grammar and by the partial consistency between its vocabulary and that of English. This is despite obvious differences between the languages, such as the order of verb and subject and the use of an inverted question mark at the beginning of questions in Spanish.

So, ease of acquiring a second language is related to the degree of consistency between new and native languages and the degree of consistency within the new language itself. The same relationship between complexity and consistency also applies to human-machine dialogues. Consistency is not the only dimension on which the complexity of a language may be judged, and later in the chapter we introduce some alternative dimensions that the designer may consider in choosing between dialogue styles. Nevertheless, since much of the research into human-computer dialogues has focused upon consistency, we adopt it as the main dimension to be discussed in this chapter.

The components of natural language

Every language has two main components: a *vocabulary*, that is, a set of words and their associated meanings; and a *grammar*, that is, rules that determine the order in which words can be used meaningfully. Many grammatical regularities are made explicit in written language by markers such as full stops, commas and question marks. Others are specified by implicit rules which dictate the order in which words may be combined, and are only recognized when they are violated (as in the sentence "man the bit dog the"). The order in which words can be combined can be determined without reference to their meaning by following *syntactic* structures in the grammar. For example, the sentence "man the bit dog the" clearly violates the syntactic rule which shows how a determiner (the) should be ordered relative to a noun (dog/man). Figure 4.1 shows how the structure of a simple sentence can be broken down into its syntactic components. The rules which describe sentence components are organized hierarchically, such that a sentence can be rewritten into its noun-phrase and verb-phrase components, a noun-phrase into its determiner and noun components, and so on, until individual words from the vocabulary match the components at the lowest level of the grammar.

The syntactic structures shown in Figure 4.1 give a very simplistic description of the grammar rules of natural language. People also make use of *semantic* structures within grammars, that is, information about the meaning

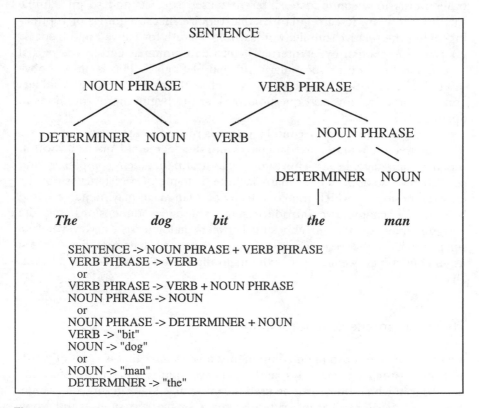

Figure 4.1 The grammatical structure of a simple sentence. The tree shows the hierarchical structure of a sentence. The rules below the tree show how a sentence might break down into its components. The -> marker signifies 'is composed of' (or in linguistics jargon 'can be parsed into').

of words and sentences. For example, in the sentence "The boy threw a brick at the window and broke it", it is clearly the window that was broken. The relative importance of syntactic and semantic information in language comprehension and production is still a matter for debate, though both are undoubtedly important. For example, in trying to make sense of the garbled sentence "man the bit dog the", one would probably use both the syntactic knowledge of appropriate orders of verbs, nouns and determiners, and also the semantic knowledge that it is much more likely for the dog to be biting the man than the other way round.

The linguistic components of machine languages

The languages of human-computer dialogues also have vocabularies and grammars, which again make use of both syntactic and semantic information. For example, consider the following program written in the programming language BASIC, which reads in four numbers and then prints their sum;

```
10    X=0: Y = 0
20    For X = 1 to 4
30    Input I
40    Y = Y + I
50    Next X
60    Print Y
```

BASIC has a vocabulary of functions (*For, Next, Input* etc.) and a number of syntactic rules which dictate how they can be used (e.g., a *Next X* statement limits the scope of a *For X..* statement), as well as a number of punctuation markers (line numbers, colons, etc.). The program also follows the semantic rule for a looping program, in which a counter is set and incremented each time the program loops. In writing programs in BASIC, the programmer has to make use of both syntactic and semantic information. Indeed, this division between syntactic and semantic information has been adopted as a model of programming knowledge by Shneiderman and Mayer (1979), a topic that will be returned to in Chapter 7 when we discuss 'programming plans' as an example of an analytical skill for understanding interfaces.

Figures 4.2 and 4.3 show two further examples of human-computer dialogues. The first example consists of commands from the operating language UNIX, issued by the user as typed word sequences in abbreviated form. The second example is a set of commands from the MacDraw graphics package in which actions are performed on screen using a direct manipulation

Examples of UNIX commands	Description of their effects
ls	(Abbreviation for "list") Lists all the files in the current directory
ls -l	(Abbreviation for "list long") Lists all the files in the current directory along with their size, date of last modification, file owner and additional file protection information
rm mbox	(Abbreviation for "remove mailbox") Deletes the file 'mbox' (a default file created for receiving electronic mail that has been read)
tr "{}" "()" < pretty.text > ugly.text	(Abbreviation for "translate") Replaces all the occurences of {} brackets in the file "pretty.text" with () brackets in a new file "ugly.text"

Figure 4.2   Examples from the UNIX command language.

interface. These radically different dialogue styles nonetheless have two things in common:

1. They both have a finite set of commands with which the user's intentions are communicated to the machine; and
2. They require the commands to be issued in ordered and restricted sequences so that the machine can interpret the commands in the way intended by the user.

The UNIX dialogue shown in Figure 4.2 is clearly linguistic in the sense that it involves interpretation by the machine of, albeit artificial, words of known meaning as a symbolic expression of the user's requirements. The UNIX vocabulary consists of commands described by artificial words, such as *ls* (list files) and *cat* (list file contents on screen). The UNIX grammar determines the orders in which command words can be combined, so that a set of related commands can be issued in one stream. Like written natural language, UNIX also requires the use of syntactic markers such as brackets, full stops and slashes to order and separate commands and file names.

The MacDraw dialogue shown in Figure 4.3 appears less appropriate for linguistic analysis at first sight, since it employs actions rather than word sequences to indicate the user's requirements. Nonetheless, dialogues such as this can be treated as analogous to linguistic structures, if action sequences are equated with word sequences. The vocabulary of MacDraw consists of a number of actions carried out using the mouse. For example, to create a square the user clicks once on a panel of icons to select the tool for drawing squares, then clicks and holds down the mouse button to activate the tool on the drawing pane, drags the mouse to draw a square of the required size, and releases the mouse button to complete the square. The grammar determines the way in which groups of actions are related, such as the actions for drawing a particular shape (point and click to select shape icon, click to locate

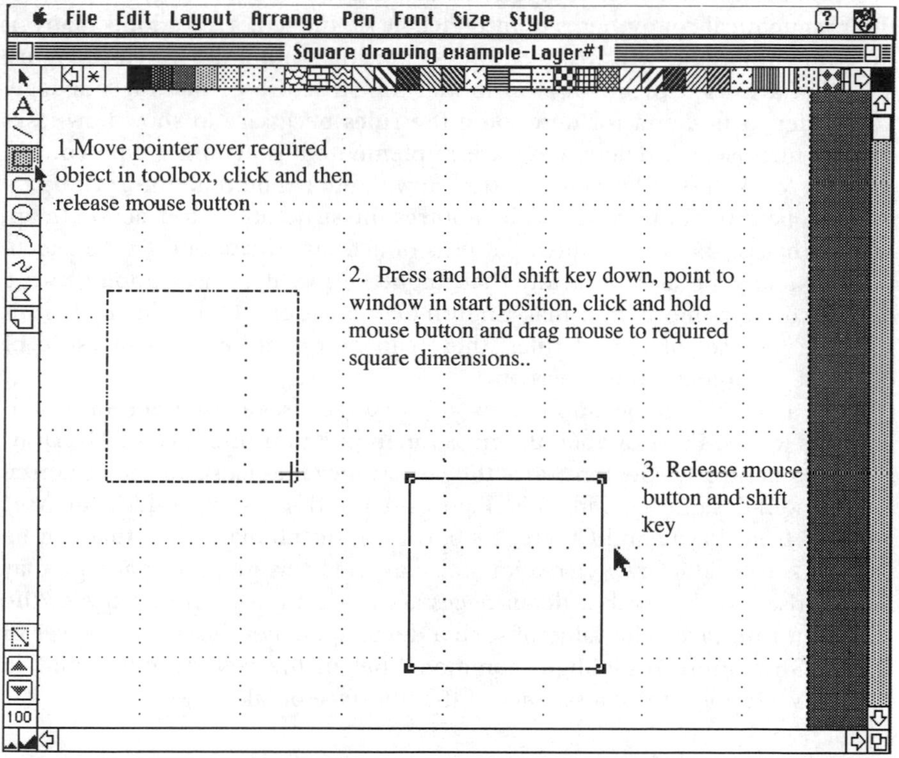

Figure 4.3 Drawing a square using MacDraw™ graphics package (the text is added to explain the actions). The shaded box in the panel of tool icons indicates that the rectangle object drawing tool has been selected in Step 1. The dotted outline of the square and the cross-shaped cursor indicate a mouse drag from top left to bottom right in Step 2. The solid lines and back dots at the corner show a finished object which is selected for further operations in Step 3.

top left, and drag to required size, then release to finish drawing shape). The grammar can be punctuated by additional actions to give specific effects (e.g., pressing the shift key whilst dragging a rectangle ensures a perfect square).

4.2   Task Action Grammar analysis

The notion of consistency can be used to predict the complexity, ease of learning and occurrence of errors in human-machine dialogues like those described above. A method for analysing interfaces that adopts consistency as an organizing principle is offered by Payne and Green's (1986) Task Action Grammar (TAG). In the same way that the syntactic rules of Figure 4.1 describe the grammatical components of a sentence, TAG analyses describe the grammatical components of interface tasks. Indeed, TAG adopts many of the same notational features (e.g., the -> parsing symbol described in Figure 4.1) that are used by linguists to model the structure of natural language. TAG offers a notation for describing the rules necessary to show how tasks (move, turn, select, delete, etc.) are implemented as actions ('type control then F', 'click and hold mouse button down', etc.). The notation is complex and can be difficult to read, but it captures the structure of task-action 'mappings', that is, the way in which patterns of actions correspond to the execution of a specific task, in a way that might be lost if these mappings were described in a more informal English-like notation. Once an analysis is reduced to a set of formal rules, these can be quantified as a measure of interface complexity and consistency.

TAG analysis can be applied to any dialogue style, whether or not it employs words. For example, the direct manipulation interface of MacDraw shown in Figure 4.3 demonstrates the consistency that exists in the organization of well-designed commands. Figures 4.4 and 4.5 show a TAG analysis, adapted from Payne and Green (1986, p.108), of the main tasks that can be undertaken in MacDraw. A reader who has used this graphics package may be surprised at the level of detail necessary to capture the properties of the MacDraw interface. The value of such a detailed analysis lies in the power of the TAG notation to highlight commonalities in the task structure, and to show at what point the consistency of the interface breaks down.

Aspects of TAG analysis

As a first step in a TAG analysis, it is necessary to classify all the tasks that may be carried out to achieve a goal set by the user of an interface. Figure 4.4 shows a list of *simple tasks* that can be carried out using MacDraw. There are basically four things that the user can do with this drawing package: they can

SIMPLE TASKS

Create object [*Effect* = create, *Case* = regular]

Create special object [*Effect* = create, *Case* = special]

Move one object [*Effect* = move, *Number* = one, *Case* = regular]

Move group of objects[*Effect* = move, *Number* = group, *Case* = regular]

Move scattered objects[*Effect* = move, *Number* = scattered, *Case* = regular]

Move one object, restricted path[*Effect* = move, *Number* = one, *Case* = special]

Move group of objects, restricted path[*Effect* = move, *Number* = group, *Case* = special]

Move scattered objects, restricted path[*Effect* = move, *Number* = scattered, *Case* = special]

Modify one object[*Effect* = modify, *Number* = one,*Case* = regular]

Modify group of objects[*Effect* = modify, *Number* = group, *Case* = regular]

Modify scattered objects[*Effect* = modify, *Number* = scattered, *Case* = regular]

Change default attributes [*Effect* = change default]

Figure 4.4   The list of simple tasks for the MacDraw™ interface (adapted from Payne and Green, 1986, with the addition of tasks for moving or modifying a group or scattered collection of objects). Each of the tasks has associated variables (shown in italics), which describe the nature of the task. The *Effect* variable stands for the user's intended task (create, move, modify or change attributes), and the *Case* variable stands for either a regular form of the task or a special case (draw squares and circles only; move objects along horizontal or vertical grid lines only). The special cases are achieved by pressing the Shift key while carrying out the mouse movements. Note that the simple task 'change default attributes' does not contain the *Case* variable, since no objects are selected and the Shift key has no effect on this task. The *Number* variable (not used in Payne and Green's TAG analysis) stands for the number of objects to be moved or modified, and can be one object, a composite group of objects or a scattered group of objects. Again, the multiple selections are achieved by pressing the Shift key while carrying out the mouse movements to select the required objects.

create an object; move an object or group of objects; modify an object or group of objects; or change a default (that is, the style that will be applied when an object is first drawn). By pressing the Shift key down during the execution of some of the actions, the user can alter these basic tasks to create only square or circular objects, to move only along vertical or horizontal grid lines, or to select a group of scattered objects. Thus twelve simple tasks are shown in Figure 4.4. The point of this description is that it does not matter whether the object that the user is manipulating is a square, circle or irregular shape: the simple task faced by the user is the same.

The next stage in a TAG analysis is to re-describe simple tasks in terms of rules for carrying out these tasks as action sequences. The rules which describe the MacDraw interface in terms of the actions necessary to undertake possible tasks are shown in Figure 4.5. The rules are effectively a grammar for generating sequences of action from simple tasks. Thus at the highest level, the MacDraw command sequence can be described by five rules that describe the general cases of using a drawing tool, changing a default style, modifying an

RULES (written in Payne & Green's notation)	ENGLISH DESCRIPTION

1.1   Task[*Effect* = create, *Case*] -> select-tool
      + point-to-2-places[*Case* , *Place1* = value-from-goal,
         Place2 = value-from-goal]
	The task of creating an object consists of selecting a tool followed by pointing to two places selected by the user.

1.2   Task[*Effect* = move, *Number* = one,*Case*] ->
      point-to-2-places[*Case* , *Place1* = value-from-goal,
         Place2 = object-location]
	The task of moving one object consists of pointing to two places, the object's current location, and the user's chosen destination.

1.3   Task[*Effect* = move, *Number* = group/scattered,*Case*] ->
      select-object(*Number*, *Case*)
      + point-to-2-places[*Case* , *Place1* = value-from-goal,
         Place2 = object-location]
	The task of moving a number of objects consists of selecting the objects, followed by pointing to two places, one of the selected objects, and the user's chosen destination.

1.4 Task[*Effect* = modify, *Number*, *Case*] -> select-object[*Case*, *Number*] + select style	The task of modifying object(s) consists of selecting the object(s) followed by a style.
1.5 Task[*Effect* = change default] -> select style	The task of changing a default style consists of selecting the required style.

1.6   point-to-2-places[*Case* = regular,*Place1*,*Place2*] ->
      action[*Kind* = point, *Place1*]
      + drag-to-place[*Place2*]
	Pointing to two places for a regular task consists of pointing to the first place and then dragging to the second place.

1.7   point-to-2-places[*Case* = special, *Place1* ,*Place2*] ->
      action[*Kind* = point, *Place1*]
      + "depress mouse button" +"depress SHIFT"
      + action[*Kind* = point, *Place2*]
      + "release mouse button" + "release SHIFT"
	Pointing to two places for a special task consists of pointing to the first place and depressing the mouse button then the shift key, then pointing to the second place and releasing the mouse button then shift key.

1.8 drag-to-place[*Place*] -> "depress mouse button" +action[*Kind* = point, *Place*] +"release mouse button"	Dragging to place consists of pressing the mouse button, pointing to the destination and releasing the mouse button.

1.9   select-object[*Number* = one,*Case*] ->
      action[*Kind* = point, *Place* = object-location],
      + "depress and then release mouse button"
	Selecting one object consists of pointing to the object location and clicking.

1.10   select-object[*Number* = group,*Case*] ->
      point-to-2-places(*Case* = regular, *Place1* = next to first
         object, *Place2* = next to last object)
	Selecting groups of objects consists of pointing to two places, next to the first object and next to the last object.

1.11   select-object[*Number* = scattered,*Case*] ->
      "depress SHIFT"
      + select-object[*Effect* = one,*Case*] ...etc.
      + "release SHIFT"
	Selecting scattered objects consists of depressing the Shift key and then selecting the first object, then the next object, and so on, and then releasing the Shift key .

1.12 select-style -> point-to-2-places[*Place1* = style menu, *Place2* = menu item]	Selecting a style consists of pointing to two places, the style menu then the menu item.
1.13 select-tool -> action[*Kind* = point, tool-icon] + "depress and then release mouse button"	Selecting a tool consists of pointing to the tool icon and clicking.

Figure 4.5 A Task Action Grammar description of the MacDraw™ interface (adapted from Payne and Green, 1986, Figure 6 page 108, with addition of Rules 1.3, 1.10 and 1.11). The left column indicates how the simple tasks shown in Figure 4.4 are redescribed as sequences of actions. The *Place* variables indicate positions for manipulating objects which are selected by the user (e.g., in Figure 4.3, *Place1* is the top left of the square and *Place2* the bottom right). The *Kind* variable stands for discrete actions carried out by the user (e.g., 'point' refers to the action of using the mouse to move the pointer to a particular location).

object, moving an object and moving more than one objects. For example, Rule 1.1 describes the general case of creating an object, and effectively states that the task of creating an object consists of the sub-tasks of selecting a tool (Rule 1.13) and pointing to two places (Rule 1.6 or 1.7). The sub-rules for selecting a tool and pointing to two places determine the type of object that will be drawn and the dimensions and regularity of shape of the object.

Extensive use is made in TAG of *variables*, and it is to capture their effects that TAG requires the restricted notation shown in Figure 4.5. Variables indicate slots which can be filled in TAG rules to indicate the particular task requirements that the user has. For example, the *Case* variable in Rule 1.1 can stand for either 'regular' or 'special', depending on whether the user wants to create an object of unrestricted dimensions (the regular case) or one of fixed radius or side length such as a square or circle (the special case). Thus the same rule can describe at a high level the creation of both regular and special objects. This allows TAG to capture the inherent consistency that exists between the actions of creating regular and special object shapes. Similarly, the *Number* variable in Rules 1.2, 1.3 and 1.4 can stand for 'single', 'group' or 'scattered', depending on whether the user wishes to move or modify a single object, a closely grouped set of objects or a scattered set of objects. The variables *Case* and *Number* allow the twelve simple tasks shown in Figure 4.4. to be described by only five main rules. Some variables also indicate values that derive from the user's actions (e.g., in Rule 1.1. the 'value-from-goal' label assigned to the *Place* variable indicates values that cannot be pre-specified but are determined by the user's mouse movements).

Like the syntactic rules of natural language shown in Figure 4.1, TAG rules are organized hierarchically. The significance of the hierarchical organization of TAG rules lies in the assumption that the size of the grammar (i.e., the number of rules necessary to describe all the task-action mappings) can be used as a metric for judging the complexity of the dialogue. Specifically, the more rules at the highest level (Rules 1.1 to 1.5) that are required to describe a dialogue, the more complex, and therefore harder to learn, the dialogue will be. MacDraw scores quite well on this index of complexity compared with many drawing packages. This is partly due to the internal consistency of the dialogue, which requires only five main rules to describe the twelve main tasks of using the interface. It is also due to the software making use of the features of the Macintosh environment (the WIMP interface, discussed in Chapter 2). In other words, the software has external consistency with other Macintosh applications.

Consistency between an interface and users' prior knowledge

So far the consideration of consistency has been restricted to the syntactic features of interface dialogues. An equally important factor in human-machine dialogues is the semantic knowledge that a user brings to learning and using the interface. Where the vocabulary of a human-machine dialogue is the same as that of natural language, one might expect that it requires less learning than entirely artificial or arbitrary vocabularies, and this is sometimes the case. However, as we shall argue in later sections, using familiar words in dialogue languages can cause confusions between real world and interface uses, which lead to error and a need to re-learn word meanings. The key issue here is maintaining consistency between the interface language and the user's prior knowledge of that language.

For example, Hypercard (for the Apple Macintosh) and Linkway (for PCs) are object-oriented systems which are often used for applications such as computer-aided learning, where the user has some freedom of choice as to what information to pursue at any stage in using the interface (the concepts of object-oriented systems and hypertext are discussed further in Chapter 8). To use a Hypercard application, the user browses through *stacks* of *cards* (essentially linked sets of displays) to move through interlinked text and graphics using mouse-driven selections. Browsing is conducted either by clicking on buttons which take one to new cards or stacks, or by selecting movement commands from a menu, as illustrated in Figure 4.6. To browse between cards within a stack using the menu, one selects either *First*, *Next*, *Previous* or *Last*, and to browse between stacks one selects from *Back*, *Home*, *Help* and *Recent*. Once the concept of browsing through cards is understood, it is fairly obvious how to use the first set of commands for browsing within a

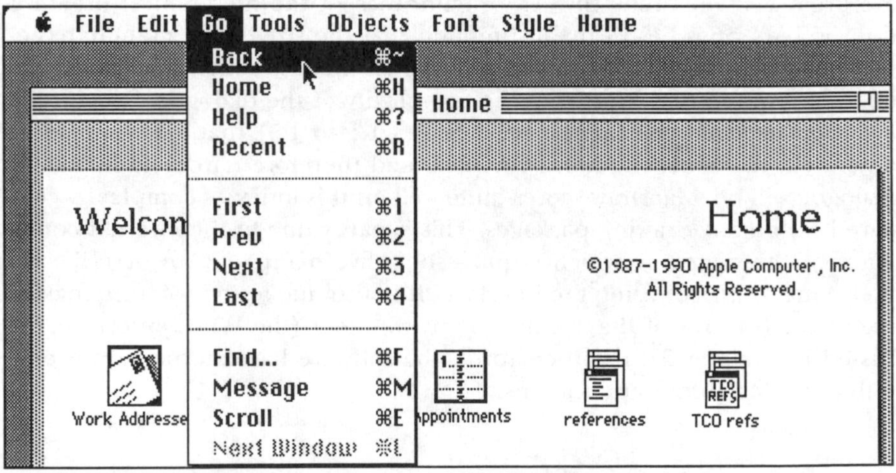

Figure 4.6 Navigation commands in Hypercard. The top four commands are for navigating between stacks, the next four are for navigating between cards within a stack, and the final four are other stack and card manipulation commands.

stack. Essentially, when a user browses through a stack they are navigating through the information it contains. The second set of commands are also words with known meanings (selecting *Back* takes you back to the last stack you browsed before the current stack, *Home* takes you to the home stack, *Help* takes you to the help stack, and *Recent* shows a reduced representation of all cards visited recently).

The navigation commands of Hypercard exemplify both the advantages and disadvantages of using command names with familiar meanings. A TAG rule to describe the selection of a Hypercard command for navigating within a stack might look like this:

Task[*Effect* = move-within-stack, *Direction*]->
    known-item(*Kind* = word, *Direction*]
    + point-to-2-places[*Place1* = GO menu, *Place2* = *Direction*].

Effectively the rule states that the task of moving within a stack consists of choosing a word that is already known and which matches the required direction, and then pointing to two places, first the GO menu, followed by the word representing the required direction in the GO menu (note the re-use from Figure 4.5 of Rule 1.6 for point-to-2-places, as an example of Macintosh interface consistency). The variable *Direction* stands for a 'known item', to use Payne and Green's terminology, that is, it matches a known word which describes the required direction of movement. There is no need to specify separate sub-rules for each direction, since the user can be taught the rule for selecting one direction and can generate the other directions from the same rule.

In navigating through Hypercard the use of familiar words to indicate directions of movement within a stack should, theoretically, reduce the number of rules that must be learned. However, it is often the case that world knowledge, especially linguistic knowledge, can have the effect of making dialogues ambiguous or under-specified. For example, our experience of teaching Hypercard suggests that the *Back* command can cause problems. Novices often have an incomplete understanding of its effects, and may ask the question "Back where?". This problem particularly manifests itself when *Back* is selected twice in succession. Does *Back* mean 'back to where you came from' or 'back even further'? The latter meaning is a correct interpretation, but novices often use *Back* in an attempt to toggle between two stacks. Part of the confusion appears to stem from the conjunction of the words *Previous* and *Back* in the same menu. Labels which offer a more accurate description of these commands are *Previous Card* and *Previous Stack*. These labels would be inconsistent with the designer's apparent decision to restrict navigation commands to single words. However, violating the consistency of command length may not necessarily be a bad thing if enhanced clarity can be gained from longer command names. Naming in command languages is explored further in Section 4.3.

Display-based information and the concept of affordance

We have argued in this chapter that consistent interfaces are less complex and therefore easier to learn. This implies that the knowledge that users hold in their heads determines their ability to perform tasks with an interface. Whilst the knowledge that a user learns is undoubtedly important, another major source of information about how to perform tasks is the interface itself, or more particularly, the *display* of the interface. Indeed, it is the effective use of displays as information sources that makes graphical user interfaces so attractive to designers, particularly in systems for novice or intermittent users.

The use of displays as information sources is illustrated in an experiment conducted by Mayes, Draper, McGregor and Oatley (1988), who examined the nature of expertise of experienced users of MacWrite, a word processing package in which commands are issued using pull-down menus. In particular, they were concerned with what expert users actually remembered of the interface itself. They found that even skilled users remembered surprisingly little of the details of the interface or the contents of the pull-down menus they were using, despite being highly competent in their use. Their conclusion was that this reflects a 'need to know' principle on the part of the users: they learn just enough to recognise (but not recall) what they want in the interface and exploit the WIMP interface to enable them to locate what they need.

Although it may seem surprising that experts cannot recall the contents of menus they use frequently and effortlessly, the explanation is simple: experts do not need to learn menus because their contents are always available for inspection. They constitute an *external memory*. The work of Mayes *et al.* challenges a justification that is sometimes cited for the guideline restricting menu lengths to approximately seven items, which is based on the need to avoid overloading short term memory. Whilst the suggested size may be indeed be reasonable, resulting from the optimization of menu search times (MacGregor and Lee, 1987 – see Chapter 2), an explanation in terms of memory limitations fails to capture the role of menus as external memory. This is not just associated with expert performance. A novice can often acquire skills with a well-designed interface simply by exploring the display. Indeed, it is possible to sit inexperienced users in front of a computer running Macwrite with the instruction that everything they need can be found by browsing through the menus, and most of them will be able to learn much of the system through exploration without further instruction (see also our discussion of minimal manuals in Chapter 6). As we shall see in the next section, the ability to learn to use an interface through exploration is closely related to the consistency of the interface.

Although consistency between perceived and actual properties of displays cannot be captured using the original TAG notation, it has been adapted by Howes and Payne (1990) to include display-based information. In the same way that Payne and Green use the notation 'known-item' to indicate that the name of a command comes from the user remembering a word that matches the intended command meaning, Howes and Payne introduce the notation 'display-item' to indicate an item that is present in the display that matches the user's required task. Figure 4.7 shows an example of this notation applied to describing the task of finding text in MacWrite. The point of this

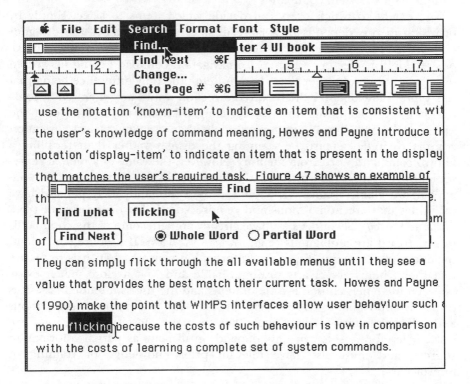

TAG rule for carrying out the task of finding a text string:

Task [*Operation* = find] ->
      action(point, display-item([*Operation* = find], [class = menu-bar]),
  + action(drag, display-item([*Operation* = find], [class = menu-bar];
                             class = pull-down, location = below;
                             class = text])
  + action(type, target string)
  + action(type, RETURN).

Figure 4.7 Carrying out the task of finding a text string in Macwrite™. The task consists of pointing at an appropriate label on the menu bar (in the case of finding text, Search is the obvious one to choose), dragging the mouse down to an appropriate text label below the menu bar and then typing the required text string and hitting the Return key. The notation 'display-item' indicates that the user looks at the display for an item that provides the best match to their current operation.

example is that the user does not need to know the names of the menu or specific menu item in order to locate the Find command. They can simply flick through all the available menus until they see a value that provides the best match their current task. Howes and Payne (1990) make the point that WIMP interfaces allow user behaviour such as menu flicking because the costs of such behaviour are low in comparison with the costs of learning a complete set of system commands.

The extent to which users can decide which items to select simply by browsing menus depends to a large degree on what names the designer chooses for the menu items. For example, the TAG rule given in the previous section for selecting a Hypercard navigation command could be written using 'display item' rather than 'known item', the only difference being that the appropriate label for the required direction would be generated by looking at the display rather than by recalling a known word. Indeed, if the Mayes *et al.* (1988) results are generalizable, then users may select navigation commands in Hypercard by scanning the display rather than recalling a known command word from memory. However, the value of a mental representation of Hypercard which relied entirely on display items would be limited, for two main reasons. First, some of the commands in Hypercard have meanings that are non-obvious and require knowledge of the system (e.g., *Home* and *Scroll*). Second, as we have already argued, some of the commands in Hypercard are ambiguous (e.g., *Previous* vs. *Back*). If users rely on menu flicking and use menu displays as information sources, then it is necessary that menu items are clearly distinguishable and that a close match exists between the labels chosen for menu items and the user's intended task. It is probably impossible to choose menu items which are entirely unambiguous and self-explanatory for anything other than the most simple of interface tasks. Therefore, users can only depend on displays as partial information sources, and must also have relevant prior knowledge or training in interface use.

Designing interfaces that use displays as information sources is a useful context in which to introduce the concept of *affordance*. Affordance is a term that appears in Gibson's (1966, 1986) theory of visual perception. Gibson argues that humans are able to recognize objects in their perceptual environment, not because of their prior knowledge about what objects should look like, but because the objects themselves have perceptual features which make their nature explicit. It would be inappropriate to go into detail here about psychological theories of perception, though it is worth noting that Gibson's theory challenges the more mainstream theory of object perception as being determined by the prior knowledge of the perceiver (e.g., Gregory, 1973). Norman (1988, p. 9) borrows the concept of affordance to describe how everyday objects such as doors and switches can be more or less easy to use according to their affordance. He states that the affordance of an object

"refers to the perceived and actual properties of the thing, primarily those fundamental properties that determine just how the thing could possibly be used."

In other words, affordance describes the extent to which the perceived properties of the 'objects' in an interface (menus, icons, pictures, indentations,
etc.) are consistent with their actual properties. Pull-down menus exhibit the
property of affordance if the names that are chosen for the commands they
contain give an accurate description of their functions.

To summarize this section, displays act as information sources, thereby
reducing the amount of information that a user must learn in order to interact with an interface. This puts the emphasis on the design of interfaces
which maximize the use of displays. For novices, this often means designing
displays that exhibit the property of affordance, that is, displays that tell the
user how the interface should be operated. Although the examples used in
this section have been about pull-down menus, the concept of affordance has
been applied to objects as diverse as doors and airplane cockpits (Norman,
1988). Interface skills which rely on exploring displays rather than remembering interface languages are discussed further in Chapters 5 and 8.

Evaluating TAG as a tool for interface design and evaluation

The previous sections show how Payne and Green's (1986) TAG can describe
the features of both text and graphics-based dialogue styles. The purpose of
such descriptions might be twofold: to develop models of the user's knowledge of interfaces; and to evaluate existing or proposed interfaces in order
to inform designers of their strengths and weaknesses. Payne and Green
stress that TAG offers only a *competence* model of user knowledge of interfaces, that is, it describes the kinds of knowledge that users *might* have of an
interface. It cannot be used to predict all the factors (forgetting, length of
response times, and so on) that will affect *performance* with an interface. As a
tool for psychologists it has its limitations, but its use as a tool for evaluators
is worthy of exploration. We return to this issue in Chapter 10.

Let us return to the example of the MacDraw interface, shown in Figures
4.3, 4.4 and 4.5. A TAG analysis captures functions of the MacDraw interface
that the designers intended to be used but that users often miss. An over-
simplified mental representation of the MacDraw interface might replace
Rules 1.1 to 1.5 with a general rule which states:

"To carry out any task, first select the appropriate item and then drag it to place."

As Payne and Green point out, there are two negative consequences of possessing this simplistic representation. The first, a relatively minor consequence, is that the user makes unnecessary mouse clicks when moving an
object, because they click on the object and then release the mouse button
in selecting the object, and then click on the object again to drag it to the

required place. The presence of the sub-task of selecting the appropriate item in the simplified rule is unnecessary for the task of moving objects, since it is duplicated in the sub-task of dragging the object to place. A second and more significant consequence is that this simplified rule does not contain any variables to indicate whether the task is a regular or special case, or whether objects are grouped or scattered. This may lead the user to suppose that the Shift key modifier only affects the creation of objects, when in fact it has a range of different effects for Rules 1.1 to 1.5. Payne and Green's admission that they only became aware of the general role played by the shift key modifier in MacDraw after performing their TAG analysis is confirmed by our own experience of the drawing package.

It is quite probable that users who learn through exploration of the Mac-Draw interface rather than by following written instructions might develop an overly simplified mental representation of the interface similar to the rule shown above, by generalizing their experience of object creation to movement, modification and default changing. Thus they would never find out about the multiple uses of the Shift key, because they would never think of using it except when they created objects. Users actively search for an economical mental representation of the rules that describe a dialogue. An impoverished mental representation can restrict the skills that users acquire with further experience of an interface. This has important implications for the acquisition of interface skills, particularly where the designer relies on exploration rather than training for developing users' skills.

One of the problems in conducting a TAG analysis to assess consistency is that alternative rule sets may represent the same tasks at various levels of detail. Indeed, in writing this section, it became clear to us that Payne and Green's (1986) TAG analysis does not capture all the tasks of the MacDraw interface. They give only six tasks (because they do not include moving or modifying more than one object) and give a smaller set of TAG rules with fewer variables to describe the tasks. In particular, they give one rule for moving objects (Rule 1.2) which only works for single objects. In effect, they ignore the role that the Shift key plays in the tasks of modifying or moving scattered or composite groups of objects. It is worth expanding on this issue because it illustrates the difficulties of using TAG as a method for modelling interfaces.

There are two ways to select more than one object for modification, depending on whether they are grouped together or are scattered across the drawing. If they are grouped together, then Rule 1.10 can be applied:

"Selecting groups of objects consists of pointing to two places, next to the first object and next to the last object."

Implementing this rule is equivalent to marking out an area of the screen, within which all objects are selected. When other objects that must not be changed lie between the objects to be modified, Rule 1.11 must be used:

"Selecting scattered objects consists of depressing the Shift key and then selecting the first object, then the next object, and so on until the last object is selected, and then releasing the Shift key."

In other words, the rule for selecting scattered objects re-uses Rule 1.9 for selecting individual objects repeatedly. Whilst re-use of Rule 1.9 may represent the understanding that a user has of modifying scattered objects, there are a number of alternative ways of achieving the same ends, such as re-using Rule 1.7, the special case for pointing to two places. The point of this seemingly pedantic critique is that TAG cannot be used to choose between psychological models of interface knowledge. The rules shown in Figure 4.5 may prescribe a way of mapping tasks onto actions, but they do not necessarily represent the knowledge structures that MacDraw users have in their heads.

TAG cannot tell us what mental representation of an interface is held by the user, but it offers a useful tool to designers for assessing interface consistency independent of the user's knowledge of the interface. However, the designer must be careful that the analysis is complete, since it is easy to suggest a spurious level of consistency with an incomplete TAG analysis. The moral of the story is that TAG is a useful tool, as long as a complete description of the simple tasks, in the form of a task analysis, is undertaken beforehand. This may seem surprising to those who have already encountered TAG in the literature, since TAG is often represented as a method of cognitive task analysis (e.g., Benyon, 1992). We would argue that TAG is a method for mapping tasks onto actions and therefore may be used as a secondary task analysis tool, but it cannot be used as the primary method for identifying tasks in itself. We will examine task analysis in the context of interface design further in Chapter 9.

4.3   Command and programming languages

Command languages vary in scale from the relatively simple (such as those of simple word processors) to the complex (such as the MSDOS operating system). Individual commands may consist of a few keystrokes (e.g., selecting the control key and the letter S at the same time to save a computer file to disk). Alternatively they might take the form of short words (e.g. *rm*, the UNIX command to delete a file), pairs of command words plus parameters (e.g., *chmod a+w filename*, a UNIX command to change the mode of file access allowing all users to have write access to the file *filename*), and even complex strings of words which issue multiple instructions.

Research into the design of command languages has investigated the effects of different vocabulary and grammar choices in command language design. The issues it raises are reviewed in following sections. We also discuss the structure of programming languages, since they share a number of common properties with command languages. Indeed it is not clear whether sophisticated command languages like UNIX might not better be described as programming languages.

There are three main properties of command languages which distinguish them from other dialogue styles (such as question-and-answer or menu-driven systems). These properties are related to the *control* of different dialogue styles, the *learning* of dialogue languages, and the *resistance to error* of different dialogue styles. A computer programming language might also be seen as a form of command language: it has a vocabulary which represent objects or attributes in the form of specific words, pictures or codes; a syntax which combines the vocabulary in a systematic way; a semantics for interpreting the meaning of a set of symbols connected by a particular syntactic form; and a pragmatics for creating programs that are efficient and easy to read. Despite these parallels with command languages, programming languages differ according to the three properties of control, learning and resistance to error.

The first property of command languages is that they are initiated and controlled by the user, and not by the machine. In command-driven dialogues, the user types in commands and the machine responds to them. Although the vocabulary and grammar of command languages are determined by the designer, when it comes to the actual interaction between user and machine it is the user who decides what to type in next and how to order a string of commands. As a consequence, command languages are very powerful expressive devices. Depending on the size and complexity of the vocabulary, the user can activate a much wider range of functions with command languages often in a much more compact and faster form than with any other dialogue style. This advantage is taken to its ultimate form with programming languages, which should be flexible enough to allow the user to issue a practically infinite number of different instructions for a machine to perform. However, programming and command languages might be seen to differ in terms of the degree of direct control: the user appears to control the machine directly with a command language, whereas control is carried out indirectly through a program with a programming language.

The second property of command languages, and a consequence of the user's control over them, is that they require learning. Most command languages use an artificial dialogue with a restricted vocabulary, which may have little or nothing to do with natural language. To exploit the flexibility of command languages, the user has to know what they want to do, what the available commands are, and in what order to issue the commands. The development of WIMP interfaces has offered an alternative to the heavy learning overheads of command languages for the developers of many computer applications. Nevertheless, the expressive power and the speed of command languages are significant advantages which will perpetuate their use. Indeed, many graphical menu systems provide a secondary command language consisting of keystroke shortcuts, which are used as the user gets more experienced and wants to make selections faster. The difference between command and programming languages in terms of learning is one of scale. Programming languages are perhaps the hardest of all interfaces to learn and to become proficient in their use. A recent reaction to this is to adopt graphical interfaces for programming, such as the development of visual programming systems like Prograph. However, it remains to be seen whether a graphical interface can really lessen the difficulties of programming significantly.

A third property of command languages is that they are error-prone. An example is to be found in the EMACS text editor. The command sequence to save text is to press the control key and X at the same time and then the letter i. On the other hand, to insert another file into the current document, one presses the control key and X at the same time and then the control key and the letter i at the same time. Operationally, the difference between these distinct functions is determined by whether one remembers to lift ones finger from the control key before pressing the letter i.

The potential for errors in using command languages can be reduced by providing appropriate feedback or by allowing errors to be corrected, issues which are dealt with in the next chapter. An important difference between programming and command languages is that, whereas one finds out almost immediately when one makes an error with a command language, errors can lie in computer programs for a long time before their effects come to light. Indeed, the evaluation of software is often based on a tolerance level of, say, five errors in every 1000 lines of code. The 'error lifetimes' of programming languages, that is, the number of attempts at debugging before errors can be located and removed, can be a useful additional measure for evaluating different languages.

Command language vocabularies

The use of technical terms, or 'computerese', is often cited as a reason why many potential users are put off computer systems. On the other hand, it may be argued that technical terms have the advantage of brevity, that technical matters require technical labels, and that 'user-friendly' language may encourage unsophisticated reasoning. Essentially, commands words should have three properties:

1. They should have clear meanings. For example, Norman (1981a) criticizes UNIX for its use of obscure labels like *pip* and *chmod*. Note that meaningfulness is not the same thing as familiarity. Indeed, it will be argued below that familiar words can sometimes make poor command names;
2. Their individual meanings should be distinctive. For example, the Hyper-

card commands *Back* and *Previous* have already been criticized for failing to signal their distinctive meanings;

3. They should be relatively easy to learn or to generate from either prior knowledge or from the display. For example, the 'vi' text editor uses the keystroke commands <control-F> and <-control-B> to move the cursor forward or backward a line, and <control-N> or <control-P> to move to the next or previous character respectively. These can be generated easily, as long as the user learns the high level rule that movement is signalled by the control key followed by the first letter of the natural language word for the required direction. An alternative but equally effective choice is the command set used in Wordstar and illustrated in Figure 4.8, which uses the relative position of letters on a QWERTY keyboard to indicate directions of movement. Howes and Payne (1990) point out that this uses the display-like properties of the keyboard as an information source.

There are other properties that well chosen command names require if they are typed at a keyboard, such as brevity and ease of spelling, but they are more a function of the input device than the linguistic properties of the com-

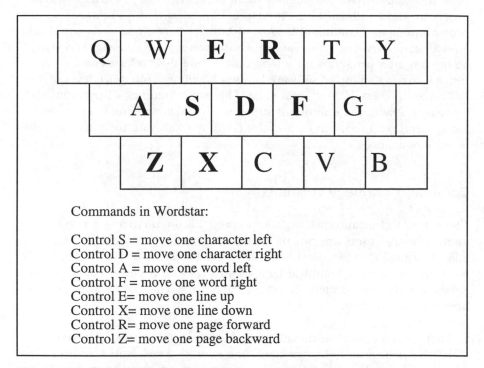

Commands in Wordstar:

Control S = move one character left
Control D = move one character right
Control A = move one word left
Control F = move one word right
Control E= move one line up
Control X= move one line down
Control R= move one page forward
Control Z= move one page backward

Figure 4.8   The command set for movement within Wordstar, word-processing software using text-based commands. Note how the direction of movement of each command corresponds with the spatial position of the command letter on the keyboard.

mand set. Unfortunately, these properties often give rise to conflicting design requirements. Meaningfulness is often achieved at the expense of distinctiveness, which is often achieved with heavy learning overheads or lack of brevity.

Perhaps the most fundamental issue associated with choosing command names is deciding to what extent the user must learn a set of command names and meanings from scratch, and how far the user can be expected to generate the command set from their own knowledge or from the display. Individual words or symbols are not always easily comprehensible on their own: they require *context*. Context is provided both by the linguistic structures in which words are used, by the situations in which they are used, and by their users. In order to offer guidance for designers on command naming, Barnard and Grudin (1988) provide an extensive review of empirical research. In reviewing studies of name generation, comprehension of name sets varying in meaningfulness and effects of different abbreviation strategies, they identify many contradictory findings. As they point out, experimental studies have frequently failed to consider the context in which commands will be used. Nonetheless, there are some findings which appear to be relatively constant across contexts.

One finding that appears to be reliable is the lack of naming consistency found in studies of user-generated command names. In a series of studies, Furnas, Gomez, Landauer and Dumais (1983) found the probability of two individuals giving a command the same name was less than 20%. Similarly low levels of agreement were found by Jorgensen, Barnard, Hammond and Clark (1983) in a study of name generation by computing professionals, and little consistency was found within the name sets of individuals. More worryingly, their subjects were very unlikely to change the names when given an opportunity to review the name set they had created. Carroll (1982a) found similar inconsistency in the names people chose for computer files. Each individual used approximately fifteen different naming strategies on average, and tended to use these inconsistently. However, computer file names serve a very different purpose from command names. It may be appropriate to have many different strategies to discriminate between file names, for example, to indicate the context of creating the file as well as its contents (Lansdale and Edmonds, 1992). There is a danger that studies of computer file naming may not generalize to computer command naming. Furthermore, the names that designers provide for their own purposes may be very different from names they would provide for naive users. Nevertheless, there is clearly enormous variation in command naming, and it cannot be assumed that either users or designers can always identify a consistent name set.

It is in studies which investigate meaningful command names that the most contradictory findings have emerged. Meaningful command names might appear to be a good idea, because individual command names can be generated from knowledge of the natural language labels for related actions (e.g., CUT, PASTE, and COPY are text editing commands which evoke knowledge about paper-based editing). Ledgard, Whiteside, Singer and Seymour (1980) demonstrated an advantage for natural command phrases over arbitrary command strings, though they varied both syntax and names across conditions. Black and Moran (1982) also found that meaningful commands can be remembered better than arbitrary commands, though they also found that less frequent words were remembered better than frequently used words, and that pseudo-words were remembered better than meaningful words which did not match the command meaning. However, Landauer, Gallotti and Hartwell (1983) found that task performance was unaffected by whether arbitrary or meaningful command names were used, although subjective ratings varied.

Rather than simply using familiar words with known meanings, it has been suggested that meaningful command names are appropriate only if they are also congruent with each other. In other words, name sets should be strongly inter-related, so that the user can generate all the other names from knowing one of them. For example, Carroll (1982a) found that congruent command name sets (advance; retreat) were easier to learn and use than incongruent sets (go; back). The two command sets used in his experiment are shown in Figure 4.9, along with the TAG rules that describe them. As the figure shows, the easier command set can be organized by one high level rule which makes use of one known item to generate the other, whereas the other command set requires two rules for each pair of commands. In Carroll's experiment the congruent commands were antonyms, though as Barnard and Grudin (1988) point out, dimensions for judging congruence can include whether words are consistently verbs or nouns, single or plural, etc. Indeed, some of the incongruent command pairs used in Carroll's experiment were not only incongruent but contained ambiguities (Go ... where?; Turn ... what way?) as a result of mixing verbs and nouns. Thus it is not clear whether Carroll's findings are a result of ambiguity or congruence. Nevertheless, both features offer important dimensions for the designer of a name set to consider.

An appraisal of experimental approaches to command naming

Research on command naming and vocabularies was prominent in the early 1980s but has petered out somewhat since. One reason for this is that the emphasis upon command languages has diminished with the introduction of graphical interfaces. The second reason why research into interface design has moved away from these issues is that they have not proven easily tractable by an experimental approach. Many of the studies in this area were of questionable generality. For example, the comparative studies of Ledgard et al. and Carroll both lay themselves open to the criticism that uncontrolled variables may have given rise to the results found. Landauer et al., on the other hand, can be criticised for testing subjects on unrealistic tasks, in which a set of only three commands was tested.

Commands	Language A	Language B
Move forward one step	ADVANCE	GO
Move backwards one step	RETREAT	BACK
Change direction to right	RIGHT	TURN
Change direction to left	LEFT	LEFT

TAG rules, Language A:

1.1 Task[*Effect* , *Direction*] -> name[*Effect* , *Direction*]

1.2 name[*Effect* = move, *Direction*] ->
            known-item[*Kind* = word, F("Advance"),*Direction*]

1.3 name[*Effect* = turn, *Direction*] ->
            known-item[*Kind* = word, F("Right"),*Direction*]

TAG rules, Language B:

1.1 Task[*Effect* , *Direction*] -> name[*Effect* , *Direction*]

1.2 name[*Effect* = move, *Direction* = forward] -> "GO"

1.3 name[*Effect* = move, *Direction* = backward] -> "BACK"

1.4 name[*Effect* = turn, *Direction* = right] -> "TURN"

1.5 name[*Effect* = turn, *Direction* = left] -> "LEFT"

Figure 4.9 Subsets of two command languages used in the experiment by Carroll (1982), with TAG rules for each language adapted from Payne and Green (1986) which account for the advantage of language A over language B. Language A requires only three high level rules to describe the commands because learning a prototypical command enables the user to generate the other as a known item (the prototypical case being indicated by the F "advance" and F "Right" notation). Language B, on the other hand, requires a TAG rule to be learnt for each command.

An example of more careful experimental design illustrates the technical difficulties faced by researchers. Barnard, Hammond, MacLean and Morton (1982) studied the learnability and memorability of command name sets as a function of the *specificity* of the terms used. For example, when adding text to the end of a document, the command *append* is seen as more specific than *add* because its range of meaning is more restricted. Many of the basic experimental measures failed to provide a significant performance benefit of specific over general command words. Differences were only detectable in the strategies of usage and other subtle indicators. This experiment suggests that naming is best achieved by avoiding words with connotations which go beyond their system-specific functionality. If this means using unusual, jargonistic, or even non-words, then so be it.

What this study also demonstrates is that experiments in this area may be rather insensitive tools. Users' performance may not be representative of normal interface use, possibly because it is measured in an unrepresentative task and over short periods of time. Further, if the experiment is based upon hypotheses concerning trends in simple measures such as response times or memorability of command words, it may fail to capture subtle aspects of user performance. Many researchers have questioned whether experiments of this type can give useful insights into important aspects of interface design. If experiments have to be much more complex, it is likely that researchers in human-computer interaction will look for other ways to approach the subject, and possibly more tractable issues.

Grammatical features of command languages

Like natural language, the grammars of command languages can be analysed in terms of their syntactic properties. Command languages can differ syntactically according to the order in which commands are typed, and the punctuation devices used to separate and annotate a command string. The sheer number of syntactic properties that a command language grammar might have make this a very difficult area in which to do research, and as a consequence we will discuss it only briefly, using one example of research into syntactic consistency.

Syntactic consistency of command languages was investigated by Barnard, Hammond, Morton and Long (1981) who looked at the ordering of commands with arguments that either recurred in every task or which appeared only infrequently. They compared performance with two sets of command-argument ordering which are illustrated in Figure 4.10, one which kept the syntactic order of arguments compatible with a natural language sentence order, the other in which the order of arguments was kept consistent with respect to the order of parameters regardless of whether the outcome was compatible with natural language. They found that subjects performed better with consistently ordered commands, especially when the parameter (the message ID) which was used in all the commands occurred first. An explanation of this effect is offered by Payne and Green (1986), who suggest that,

Command	Natural Order	Consistent order
"Delete [string] from the file [messageID]"	DEL [string] [messageID]	DEL [messageID] [string]
"Store [messageID] in location [folder]	STO [messageID] [folder]	STO [messageID] [folder]

Figure 4.10 Command orderings from Barnard *et al.* (1981). The natural command order keeps the direct object of the sentence before the indirect object, whereas the consistent order keeps the recurrent argument (i.e., an argument that occurs in every command) first.

although more TAG rules in total are required to describe the consistent ordering, there are fewer high level rules to learn than with the natural ordering.

Whilst it seems reasonable that maintaining a consistent syntactic order might outweigh any advantages that compatibility with natural language may confer, there are problems with this research. Barnard *et al.* found the advantage for consistent ordering was not as great when the parameter used in all commands occurred second. A subsequent experiment by Barnard and Hammond (1982), in which subjects had to remember command names rather than having them presented as in Barnard *et al.*'s (1981) study, did not find the same ordering effects. This suggests that the differences between orderings observed in the earlier experiment may have been the result of subject strategies specific to that experiment. However, this does not change the overall conclusion that consistent syntax is a desirable property of command languages.

Notwithstanding the results of Barnard *et al.* (1981), natural ordering may still be an important factor in designing interface languages. For example, Ormerod, Manktelow and Jones (1993) found that natural language ordering was a significant factor in the comprehension, rephrasing and writing of conditional rules. To summarize their findings, subjects' ability to reason with different rule forms was affected by both the syntactic ordering of the rules and the temporal relationship between rule components. Similar effects may account for some errors in programming. Whilst programming languages such as Pascal require the structure "If X then Y", conditional rules must be structured in the language Prolog as "Y if X". The conditional rule form of Prolog can be a source of difficulty, particularly for programmers who have experience with other languages. For example, a common error is to express rules such as "if X > 5 then Y = 0" as "X > 5 if Y = 0" when writing them in Prolog's notation, which puts the antecedent (X) and consequent (Y) in the wrong logical order.

4.4   Dimensions for dialogue design

The limitations of consistency as a dimension

So far, our linguistic analysis of dialogues has focused on the notion of consistency. However, consistency is only one of the dimensions on which a dialogue might be judged. Indeed, aiming for consistency at the expense of all other aspects of interface design can have unexpected and unwelcome consequences.

The problems of focussing only on consistency are illustrated by Young's (1981, 1983) description of the input formats used in three different styles of hand-held calculator: a Reverse Polish Notation (RPN) calculator in which the user enters calculations according to a postfix format, the operators coming after the arguments, such that the sum $2 \times (3 + 4)$ is entered using the key strokes "2 Enter 3 Enter 4 + ×"; a four function (FF) register calculator which has the normal infix arithmetic notation, but does not have brackets, such that the user must enter "$3 + 4 = \times 2 =$" ; and a full algebraic (ALG) calculator which has infix notation and provides brackets, the user entering "$2 \times (3 + 4) =$".

The three calculators each have advantages and disadvantages. Not surprisingly, the RPN calculator is hard to use unless one becomes familiar with Reverse Polish Notation and the stack registers around which it works. The ALG calculator is entirely consistent with well-learned rules of arithmetic expression, and is generally found to be easy to use because of the simple transfer of skills. Interestingly, as Norman (1983) points out, although the FF calculator is very easy to learn, users often fail to exploit the full power of the device by using only those functions which are consistent with their prior skills, ignoring the more powerful functions.

Use of the ALG calculator illustrates how users will transfer their existing arithmetic skills directly when the interface allows it. However, the example also illustrates a potential disadvantage of transfer. In the case of the FF calculator, the implicit knowledge of operator precedence in the arithmetic expression $2 \times 3 + 4$ cannot be captured directly by the interface, which requires the order of keying in to be changed, and is often a source of errors with novice calculator users. Even though the ALG requires more notation and key strokes, it is a much simpler calculator to use than the FF. Interestingly, the RPN is harder to use than the ALG, but people who take the trouble to master it often say that it is much more efficient in the long run. It is therefore sometimes the case that a little investment in training will produce greater efficiency.

Consistency between new interfaces and old skills needs careful consideration. In considering the internal consistency of interfaces, learnability and consistency are closely related: the more internally consistent an interface is, the less there is to learn. Using common functions in different application programs, as in WIMP interfaces, is a way to make interfaces easier to learn and use because they enhance transfer of skills. The problem with consistency as a topic is that, while it represents a generally desirable objective of interface design, it is rather hard to specify exactly what has to be consistent with what (see Grudin, 1989). Nevertheless, consistency has been central to many philosophies of design which have found themselves expressed as dialogue design guidelines. It is not surprising that such guidelines sometimes require trade-offs between inconsistent or contradictory recommendations (Maguire, 1982).

Alternative dimensions on which to judge interfaces

If the dimension of consistency on its own is insufficient to characterize the structure of interfaces, then what others might be examined? Green (1989) describes a number of other dimensions which can be used to evaluate properties of interfaces, some of which are discussed below. It is fair to say that Green's dimensions are particularly appropriate for interfaces which will be used creatively by designers (e.g., programming languages or computer-aided design systems), and are less applicable to simple command-driven interfaces like automatic bank tellers and simple word processors. Green's main thesis is that interfaces which are to be used by designers need to take account of the way in which designers carry out the process of design.

Dependencies

Many complex interfaces, such as spreadsheets and programming languages, involve dialogues in which the user sets up a calculation in one part of the dialogue that depends on the results of another calculation elsewhere in the dialogue. For example, Figure 4.11 shows an example of a Microsoft Excel spreadsheet to carry out various statistical calculations. As Green (1989) points out, spreadsheets typically show only the local dependencies that exist between cells, by allowing the formula for each cell to be displayed in turn. There is usually no mechanism for examining the long-range dependencies that may exist between cells (e.g., between cells B24 and C75 in the example shown in Figure 4.11). One consequence of this is that experienced spreadsheet users spend much of their time inspecting the formulae of cells simply to unmask the hidden dependencies that exist. The challenge for designers is to provide tools alongside an interface that unmask hidden dependencies, such as browsers for examining many cell formulae simultaneously with spreadsheets, or program tracers with programming languages. The disadvantage of providing such tools is that they can add to the size and complexity of the interface.

 Fi͒ Edit Formula Format Data Options Macro Window   ⑦ ⚔

| Normal   ⬇ | ← → ⊞☰ Σ **B** *I* ≣≣≣ ☐ ╲☐◯◝ ⣿☰◻◉ |

| C75 | =SUM((C32*C32)+(E32*E32)) |

≣≣≣≣≣≣≣≣≣≣≣≣≣≣ 2 way ANOVA e.g. ≣≣≣≣≣≣≣≣≣≣≣≣≣

	A	B	C	D	E	T	U	V
13								
14	Table 35a: AB (similarity x proximity) summary table							
15		A1 (similar)		A2 (dissimilar)				
16	B1 (close)	86.0		129.0				
17		60.0		95.0				
18		79.0		86.0				
19		45.0		102.0				
20		36.0		60.0				
21		61.0		79.0				
22		51.0		74.0				
23		**58.0**	**476.0**	**95.0**	**720.0**			
24	B2 (Distant)	97.0		155.0				
25		69.0		97.0				
26		88.0		94.0				
27		63.0		119.0				
28		52.0		58.0				
29		65.0		88.0				
30		96.0		90.0				
31		**72.0**	**602.0**	**95.0**	**796.0**			
32	Totals A (Ta)		**1078.0**		**1516.0**			
33								
53								
74								
75	∑Ta2 (similarity) =		**3460340.0**					
100								
101								

Cell formulae:

C75 = SUM((C32 * C32) + (E32 * E32))
C32 = SUM(C31 + C23)
C31 = SUM(B24:B31) etc...

Figure 4.11 Dependencies in a Microsoft Excel spreadsheet. Note the formula bar at the top displays the formula of only one cell at a time, hence the need to list dependent cells below. An error in one of the cells B24 to B31 would propogate through to cell C75 giving the error message #VALUE!, but would not show the link between the cells, thereby making it difficult to debug the spreadsheet formulae.

Viscosity

A notation that is viscous is resistant to local change. A viscous notation will be more difficult to use because any changes that are made at a later stage in interface use will affect decisions made earlier. For example, in writing this book we have constantly had to change the numbering of tables and figures, and so on, as our plan of the book develops. Maintaining cross-references within the book is therefore quite a difficult task. The word-processing software we use helps to some extent by offering a facility for automatically updating section numbers, but it does not remove the problem entirely. As well as providing automatic updating methods to propagate changes through a system, viscosity can be ameliorated by making interface languages modular. This is particularly the case with programming languages, where a modular style is encouraged (i.e., writing programs in small chunks which can be tested independently and do not have dependencies between each other). Avoiding viscosity is, therefore, partly a requirement for the notation but also can be controlled by using the notation appropriately.

Hard mental operations

Another dimension of interface notations is whether a notation forces the user to undertake any particularly hard mental operations simply to comprehend the notation itself. Many of these operations are a result of the need to translate complex grammatical expressions. For example, Manktelow and Jones (1987) describe the following sentence found in a polytechnic prospectus, which had their colleagues baffled as to how it should be interpreted;

> "Do not use this form if you are applying only for courses other than first degree or DipHE."

As they point out, the negatives used in this sentence make it extremely difficult to identify what a student should do.

The difficulties of dealing with negation are illustrated in an experiment by Green (1977), who examined the comprehension of different kinds of 'looping' notation by professional programmers. Programs were constructed using three notations to direct the flow of control: a JUMP notation, which used an *if... then goto ...* command; a NEST-Begin-End notation, which used an *if... then ... else ...* command; and a NEST-If-Not-End notation, which used an *if... then ... if not ...then ...* command. The final notation contains a negative statement, which is essentially unnecessary for determining the flow of control. Green found that both of the NEST notations were better than the JUMP notation for tasks where the programmer had to identify what happened next in the flow of control of a program. This is equivalent to the task faced in writing a sequence of programming steps. Additionally, the second NEST notation was better than the first NEST notation where tasks involved identifying the conditions that led up to a certain point in a program's execution. This task is equivalent to that faced in trying to debug a program. Thus, despite the fact that the second NEST notation was more syntactically complex than the first, it provided an explicit negation which programmers otherwise had to make implicitly in their heads. This experiment provides an example of how some programming notations introduce unnecessarily hard mental operations. They also illustrate how the provision by a designer of an appropriate addition to the notation (in this case, an essentially redundant negative statement) can remove the burden of this hard mental operation.

Role expressiveness

This dimension describes the extent to which an interface language signals the meaning and function of dialogues sufficiently (e.g., in the way that the *For . . . Next, While . . . do . . .* and *Repeat . . . Until* looping notation of the Pascal programming language signals the type of program that is being studied).

Premature commitment

Some interface languages allow the user to postpone decisions that would fix the nature of the dialogue before it has been properly planned (e.g., in the object-oriented programming language Smalltalk 80, the inheritance hierarchy must be constructed from the highest level down, making it very hard to change a program design once it is started). Thus, the designer of interface languages must aim to minimize the amount of commitment that a user makes in early stages of a dialogue.

4.5 Natural and artificial dialogues

Imagine being able to instruct your video recorder as follows: 'I would like you to record the 3pm football game on Channel 4 and then the 9.30pm film on BBC1, and make allowances for any changes in the schedule that might occur.' Natural language interfaces are a very attractive proposition, since they do not require the learning of a new notation. Ideally, one might use the same language in communicating instructions to a machine as one would to another person.

However attractive the prospect may seem, there are a number of problems inherent in natural language dialogues, of which five potential difficulties are illustrated in Figure 4.12. Some of these reflect the inadequate understanding we have of natural language itself. The development of automated natural language systems is still fairly primitive, due to the enormous *complexity* of natural language. An interface capable of dealing with true natural language would have to handle an enormous number of possible natural inputs, and can therefore be considered beyond the scope of most interface designers, at least for the foreseeable future (for a review of natural language generation for interface designers, see Mykowiecka, 1991). Also, as the example given above of pre-programming a video suggests, natural language dialogues can be verbose. Even if reliable speech input devices can be constructed the problems of complexity in natural language understanding will remain. Therefore, the brevity of artificial as opposed to natural language dialogues will remain an important consideration.

Even if we were able to supply sophisticated natural language understanding mechanisms for computers, there would still be drawbacks to its use as the language of interface dialogues. As the ambiguity in the 'shampoo' example of Figure 4.12 suggests, natural language lacks sufficient power to

Complexity:	Natural language is exceptionally complex and subtle. Consider, for example, the difficulty of getting a machine to understand what 'it' refers to in these sentences: John threw a brick at the window and broke it. John threw an egg at the window and broke it. John threw an egg into the bowl and beat it. John threw a glance at the window and beat it.
Ambiguity:	Natural language comprehension often requires implicit assumptions to be made. An example used in teaching programming (Arbib, 1977) is the following instructions on a shampoo bottle: "To use, wet hair, apply shampoo, rinse and repeat". One imagines a poor soul trapped in the shower in an endless cycle of hair washing.
Structure:	Natural language does not always provide the best notation for complex interface tasks, because tasks often require specialist notations (e.g., music, algebra). For example, in pre-programming a video, many people would not consider schedule changes in issuing natural language instructions to their video system. Even if they did, it does not provide a mechanism for avoiding the problems of rescheduling. A dialogue for video programming should support this task feature explicitly.
Expectations:	Natural language dialogues might lead the user to assume understanding by the machine. Weizenbaum (1976) set out to illustrate the dangers of this in his program ELIZA (see Figure 4.13). People make assumptions when using seemingly natural programming languages. For example, Taylor and Du Boulay (1986) suggest that many novice errors in Prolog programming occur because of the spurious naturalness of its notation.
Verbosity	People can afford to be verbose in human-human conversation, because human speech is a fast medium for delivering instructions, but this is problematic for typed dialogues. Interfaces employing speech decoding mechanisms offer a possible solution, though Helander, Moody & Joost (1990) suggest only 90% reliability for current speech recognition systems.

Figure 4.12 Some of the problems of using natural language as a dialogue style for computer interfaces.

express important information, simply because it is implicit in normal con-versation. For example, natural language does not often require stopping functions, that is, explicit statements of the duration of commands, because people have an intuitive understanding of what duration is reasonable. Com-mand languages need them because a computer has no intuitive understand-ing of duration. Another important point is that dialogues should provide a *structure* for thinking about complex tasks. Users do not always conduct dia-logues with computers simply to impart perfectly thought out instructions. Many tasks (e.g., programming, computer-aided design) require dialogues to support creative thinking. Finally, natural language creates *expectations* about the extent to which shared understanding exists between the user and the computer. If it were possible to communicate with a machine in natural lan-guage, it might well lead the user to assume that the machine understood a lot more that it in fact does.

Research into natural language systems

Despite the problems of using natural language for computer interfaces, there have been a number of attempts to use artificial intelligence methods to develop interfaces which are capable of understanding instructions given in natural language. For example, the program ELIZA devised by Weizenbaum (1976) uses a simple interaction in seemingly natural language to simulate a psychotherapist responding to a client's worries. An example of an interaction between ELIZA and a user is given in Figure 4.13. Eliza is ulti-mately flawed in its ability to maintain the dialogue, as the example in Figure 4.13 illustrates, because the system's 'comprehension' of the user's input is based on a simple device of syntactic and key-word pattern-matching. The dialogue breaks down whenever the semantic content of the dialogue requires a more sophisticated response from the machine than pattern matching can generate.

Natural language interfaces have been developed for computerized tasks such as computer programming. For example, Biermann, Ballard and Sigman (1983) found that novices' performance in solving algebraic equa-tions using NLC, a natural language programming system, was at least com-parable with, and often better than, PL/C, a programming language with a highly restricted syntax, even though their subjects received considerably more training in PL/C. Research has also studied the use of query languages, that is, specialized languages for posing questions in order to extract infor-mation from database systems. Capindale and Crawford (1990) review a number of studies which compare performance using formal query lan-guages (e.g., SQL, or Structured Query Language, a restricted syntactic form for database querying which is becoming accepted as an industry standard) with a number of natural query languages. Whilst some studies (e.g., Kelley,

User:	"My mother hates me".
Eliza:	WHY DO YOU SAY THAT?
User:	"She prefers my sister".
Eliza:	TELL ME MORE ABOUT YOUR SISTER.
User:	"My sister hates me as well"
Eliza:	**WHAT MAKES YOU THINK THAT YOUR SISTER HATES YOU AS WELL?**
User:	"She pulls my hair"
Eliza:	TELL ME MORE ABOUT YOUR HAIR.

Figure 4.13   An example of a dialogue with ELIZA (after Weizenbaum, 1976). The user types in sentences, which are then pattern-matched against a set of template sentences and keywords, and a response is generated containing some of the words of the input sentence (e.g., "She X my Y" would generate the reply "Tell me more about your Y").

1983) report good performance on simple and familiar data retrieval tasks using a natural query language, other studies have found no advantage for natural query languages over formal languages, and have even suggested that sometimes users prefer formal query languages (Jarke, Turner, Stohr, Vassilou, White and Michielsen, 1985).

The main problem with using natural language for querying databases seems to be that people construct queries in which the description of the required items is too loose to be interpreted by the computer. Capindale and Crawford (1990) suggest that the utility of natural query languages can be enhanced if the system provides *feedback* on the interpretation placed by the computer upon the user's queries. In this way, any ambiguities in user queries can be clarified before undertaking a database search. They describe a study of causal users' performance with INTELLECT, a natural query language which echoes an interpretation of user queries before searching the database for a solution. Their observations suggested that the language was both efficient and well received, though no comparison with performance using a formal query language was offered.

An alternative solution to the problem of under-specified natural language queries is offered by Zoltan-Ford (1991). She argues that the problem with user queries phrased in natural language is not one of ambiguity, but that there are too many alternative ways of expressing the same request in natural language. In an experimental study she demonstrated that users could be 'shaped' by the provision of restricted computer prompts. After conducting a number of dialogues with a system, users began to restrict the ways in which they expressed queries to the language they had seen echoed by the system, notably using terser phrasing. In other words, users learned to limit their own forms of expression to those they knew the machine could deal with. What is not clear, however, is whether allowing users to acquire a restricted form of natural language through experience of a system is better than training them with a formal or artificial query language in the first place.

4.6   Conclusions

This chapter has looked at the linguistic nature of human-computer dialogues, particularly in terms of the components that represent the vocabulary and grammar of interface languages. Task-Action Grammar analysis offers a framework for analysing the linguistic properties of dialogues, and can be useful in the design and evaluation of interfaces, as long as a thorough task analysis is conducted. One of the key concepts emerging from this kind of analysis is that of consistency. Essentially, the greater the consistency both within a dialogue itself, and between the dialogue and existing user knowledge, then the easier it will be to learn and use at an interface. Consistency is not the only dimension on which the properties of a dialogue might be judged, however, especially when designing languages for undertaking complex and creative tasks. Finally, we have considered the potential for using some kind of natural language as an interface dialogue. The evidence for and against the use of natural language is, to say the least, equivocal. As a general rule, though, it appears that the more complex the task to be carried out at the interface, the less appropriate is the use of natural language as a dialogue style.

In assessing the relative advantages of artificial and natural languages, a key feature that emerges is the importance of *feedback*. The provision of appropriate feedback can clear up ambiguities in natural language expressions, it can let users know when errors are made with artificial languages, and it can help the user to gather information about the dialogue from the display that the interface presents. In the next chapter we take a closer look at the role of feedback, and we examine how it can both confirm and correct the use of languages for conducting human-computer dialogues.

CHAPTER 5

Maintaining dialogues

Overview

This chapter is concerned with the maintenance and control of human-machine dialogues by the user. Users require considerable and diverse feedback from interfaces to understand and control them. This requirement goes beyond mere knowledge of the results of actions and commands. It includes feedback which is technically redundant in the sense that the information it provides is either available elsewhere or is not strictly necessary for the job at hand. Therefore, in interface design, we need to consider the difference between what is *logically sufficient* in design to enable an interface to be controlled, and what is *psychologically necessary* for the user to experience that control.

In this chapter we focus upon three qualities of an interface that affect the the extent to which a user has control: its predictability; the feedback provided to the user; and the ability to reverse operations or correct errors.

5.1 What do we mean by control?

We have already argued in the preface to Part 2 that human-human dialogue is peppered with mechanisms which facilitate smooth conversation. Mechanisms of control, such as the use of prosodic cues (stresses and intonations used to add emphasis), indicate when it is time for the other speaker to take the floor. Normally, conversation is so smooth, or is repaired so easily, that this 'control' of human-human dialogue is automatic and hard to see. However, small examples of the breakdown in human-human dialogue can be illuminating. For example, Beattie, Cutler and Pearson (1982) illustrate how Margaret Thatcher's media interviews were characterized by interruptions, particularly ones which failed to produce a change in turn. That is, the interviewers attempted to interrupt but she ignored them and carried on speaking. False interruptions in Thatcher's interviews resulted from the mistaken belief by the interviewer that she intended to end her turn. This can be explained by the observation that she sometimes used turn-yielding cues inappropriately, and she sounded as if she were about to finish when in fact she was not. While not resulting in total breakdown of the interview, these cues caused uncertainty in the minds of the interviewers because of their inability to steer the course of conversation, and therefore diminished their ability to control the flow of the interview.

Breakdowns also occur in human-machine communication because of inappropriate interpretation of system feedback by the user. For example, when using an electronic mail system at a remote terminal, response times for echoing typed characters to the screen can be slowed down when the associated mainframe computer is being used heavily. A user may type a set of characters, fail to see them echoed on the screen instantly, and so re-type them in the belief that the system had somehow failed to register their original typing. Similarly, Payne (1991) describes how unnecessary delays in the use of automatic bank teller machines can arise from a combination of erroneous beliefs of the user and absence of the right kind of feedback from the interface. Even regular users can be seen to wait for some time between issuing each set of instructions (code number, choice of service, amount required etc.), because they believe that the system messages indicate when the machine has finished the last command and is ready for the next, when in fact the system messages simply act as prompts for inexperienced users.

These examples from human-human dialogue and human-machine communication illustrate the three aspects of control which we emphasize in this chapter. First, there is the use of prior knowledge to *predict* the likely course of development from the current status and recent patterns of a dialogue. In the example of the Thatcher interviews, prosodic cues were important in predicting turn-taking in conversation. In the electronic mail and automated teller examples, users clearly have expectations (potentially erroneous) about the speed at which the dialogues should develop. Second, there is the use of *feedback* from the dialogue to confirm (or deny) that the predictions being made still apply. In our examples, Thatcher's continued speaking, the absence of immediate echoing of typed letters, and the expected prompts of the automated teller system provided a kind of feedback as to the course of the dialogue. Finally, there is a process of *repair* to reverse or correct the course of dialogue when these predictions are not met. This would be achieved by remarks from Thatcher such as "If you would be so good as to let me finish my point ...", by the user deleting the unwanted characters at the electronic mail interface, or by the user introducing delays between commands at the automatic bank teller.

Before moving on to discussing the three topics of prediction, feedback and repair, it is interesting to see how they correspond to the practitioner's approach to design. For example, Cole, Lansdale and Christie (1985; see also Baecker and Small, 1990) suggest that a user must be able to answer the following kinds of questions in order to feel in control of a dialogue with a machine:

Where am I? In the same way that people appear to model the intentions of those they speak with, a user needs to understand the state of the machine they are working with and to predict its responses to a range of possible actions. Feedback provides this information;

How did I get here? If an interface changes in some way as a result of a particular action, it is natural for us to assume a causal link between the action and that change. If people cannot establish a causal link, either because they do not know where they are or because they do not know what they did to arrive there, then learning is impossible. Again, this information should be provided by feedback;

What can I do? Understanding the current state of a system involves knowing what tasks are possible and how they are carried out. This includes the ability to predict responses to inputs. Equally, the act of repair involves knowing what to do to remedy a current situation;

Where can I go next? Control requires knowing how the dialogue can be manipulated to enable other tasks which are not possible in the current state of the system. The user must know how to navigate around the system. This also is mediated by feedback and the ability to predict outcomes.

Answers to these questions should be seen as rules of thumb rather than formal rules. Their purpose is to alert designers and evaluators to user needs which they may have overlooked. They illustrate the ways in which prediction, feedback, and repair are needed for control, and the complexity with which these factors are interrelated.

5.2   Control through prediction

Prediction and interface consistency

Control that is based upon the ability to predict the status of the system is advantageous because the user knows when actions are or are not critical and what aspects of the interface are relevant to those actions. Therefore, monitoring of the interface status is reduced to relevant cues only, reducing the attentional demands upon the user. An example of predictability was discussed in Chapter 2, where colour codings elicited prior associations which could be used to predict relationships between, for example, RED and safety levels. Prediction was also discussed in Chapter 4, where the notion of Task-Action Grammars (Payne and Green, 1986) was used to describe how the complexity of an interface can be assessed as a function of *consistency*. There is clearly a strong relationship between consistency and predictability, since a consistent dialogue is by implication a predictable one. Although the issue of consistency has been discussed in the previous chapter, there are some additional observations to be made in relation to control of dialogues through prediction.

We need to distinguish two senses in which users of interfaces can predict the outcome of an action. They can either *know* what is about to happen, or they can *guess* intelligently on the basis of what else they know. What we know and understand is often seen in terms of training, and raises questions such as how much has been learned, and how long it has taken to learn. We can therefore talk about interfaces in terms of their 'learnability'. Guessing successfully, on the other hand, requires that users recognize situations as being similar to other situations, and that valid actions are of the same form in both circumstances. This is where we need to consider the *consistency* of the interface and the ability of the user to *transfer* their skills from one situation to another.

Prediction and users' models of interfaces and machines

Underlying the use of prediction in maintaining interface dialogues is the notion of the user having a *model* of the interface: just as we model conversations to predict appropriate contributions, interpret feedback from the other speaker, and initiate repair when conversations break down, so it is clear that users adopt *models* of interfaces to mediate their control of them. By 'model', we mean a simplified representation of the interface that strips away the complexity of the system to leave the main concepts. For example, Gentner and Gentner (1983) describe two alternative simplified models of electricity flow; a water flow model and a moving crowd model. Each of these models captures some but not all of the features of electricity, and each is a useful explanatory device.

In Chapter 1, a number of different views of an interface were described in Figure 1.3, adapted from Young (1983). Each of these views, from the perspective of the user, designer, device or psychologist, is in effect an opportunity to form a model of how the object being viewed operates. Thus, the user has a model of the device, the designer has a model of the user, the psychologist has a model of the user, the device can reflect a model of the user, and so on. Of particular relevance to the control of interfaces is the user's *mental model* of the interface and the machine, that is, the simplified mental representation that a user has of the machine operation and the effects of interface dialogues.

Psychologists are interested in the study of mental models as a way of studying how the mind works (e.g., see Gentner and Stevens, 1983, and Rogers, Rutherford and Bibby, 1992, as examples of work separated by a decade). The study of mental models has been influential in studies of users learning new interface tasks, and is discussed further in Chapter 7. In explaining the control of interfaces, we would argue that the concept of mental models contributes little more than reminding us that users use their (sometimes inadequate) knowledge to make predictions about how an interface will work, and seek feedback to confirm it (Lansdale, 1985). For this reason, in this chapter we concentrate upon the qualities of interfaces, in terms of prediction, feedback and repair, without worrying unduly about how the relevant knowledge is represented in the mind.

5.3   Control through feedback

Answering questions such as 'Where am I?' or 'What did I do to get here?' requires information. Additionally, since people learn by relating consequences to their actions, they require immediate feedback because delay will de-couple the sense of cause and effect (Michotte, 1963). This turns out to be a very important issue for interface design, because *any* observable behaviour of the interface can be seen as offering 'feedback' and users need a great deal of it, all of the time, to maintain control of the system.

Feedback is a simple engineering notion in which a system uses information about the changing environment to regulate its further behaviour. For computer-controlled processes such as robotic machine tools, the relationship between feedback and change in the system must be controlled by a programmed algorithm. Computers handle complex relationships by the use of condition-action programs such as '**if** X is present **and** Y is true **then do** action Z **otherwise do** W'. In this way, a clear programmable link between feedback and action can be established.

Human use of feedback is more complex than this, because the relationship between the environment and human behaviour is rarely based upon simple rules. We are rarely in a position to say unequivocally of the world that it is adequately described by statements such as 'X is present and Y is true'. As a result, humans create simplified models of their environment, where we limit our interpretation of the events around us to a mental representation of what is taking place and what in the environment is relevant. Feedback serves two complementary functions in supporting this process. The first is to *confirm* to the user that their model is appropriate, thereby reinforcing the user in their belief that certain data are relevant and hence aiding the efficiency of their processing. The second is to *negate* the users' model of the world, either by straight denial of expectations or by focusing attention on aspects previously ignored.

The efficient use of feedback in interfaces presupposes that feedback must be adequate but not intrusive. On the other hand, some types of feedback must force themselves upon the user's attention, either to warn of a potential error or system malfunction, or to notify the user of extraneous information which requires their attention, such as an alarm state. Design in this case is directly aimed at intruding upon the user's attention and ensuring the correct interpretation of what is happening. It is a mistake to think that feedback should be minimized to limit the information the user is asked to process. More likely, the availability of diverse sources of feedback, most of it strictly irrelevant to the job in hand, simply means the user has the information when required. This need not cause information 'overload' if done sensibly. In the following sections we deal with confirmatory or positive feedback which maintains dialogues, and we discuss negative feedback which notifies the need to take action to avoid, or recover from, errors.

5.4   Feedback to reinforce users' behaviour

Users extract information from an interface to confirm their control over it. Ironically, the effectiveness with which users interpret confirmatory feedback is often not apparent until it breaks down and the user makes errors or loses control over the system. When this happens, users receive negative feedback in the form of error messages, alarms, and unexpected actions. As in the maintenance of human-human dialogue, users require constant useful feedback throughout a human-computer dialogue. The requirement for confirmatory feedback has wide implications for the design of a system interface, and designers should never underestimate how important the provision of apparently redundant feedback can be. First we look at how confirmatory feedback acts to make the functionality of interfaces apparent to the user. This is then illustrated by examining feedback from visual displays, auditory feedback and response times as a form of feedback.

Confirmatory feedback is related to the concept of *affordance* which we introduced in Chapter 4. If an interface has the property of affordance, then its utility and functionality are self-explanatory to the user. The provision of confirmatory feedback gives the designer a mechanism for enhancing the affordance of an interface. As Gaver (1991) has argued "when affordances are perceptible, they offer a direct link between perception and action; hidden and false affordances lead to mistakes". One view of feedback, therefore, is that makes the properties of the objects we use explicit. This is the main thesis of Norman's analysis of the everyday objects that we use and misuse (Norman, 1988). What is interesting about this notion is its increased emphasis upon *design* and its reduced emphasis upon *learning*.

An example of enhancing affordance by the addition of confirmatory feedback is the use of animated icons (Baecker, Small and Mander, 1991). Animated icons display their function by providing a 10 to 20 second sequence which depicts their effects. Baecker *et al.* found that animated icons were better than their static equivalent (depending upon how accurate the animation was) at indicating the purpose of the icon (i.e., the function it would activate) and how it differed from other icons. One reason for this would seem to be the enhanced feedback which the system is providing to the user and the commensurate reduction in the amount that users have to remember about the functionality of the icons. The clarity of the animations, particularly in the depiction of sequential activities, adds affordance in the sense that the functionality is clearer and the requirement for prior knowledge is reduced.

Feedback from visual displays

The majority of human-computer interfaces involve visual displays. These usually take the form of VDUs, although many interfaces involve other devices such as the switches and dials of process control panels. In Chapter 3 the ease with which users extract information from displays was discussed, and we saw how important it is to consider the user's perception of structure in information, as indicated by format, coding and other representational devices. Errors that result from discrepancies between the perceived and actual structure of information can be avoided by giving the user more appropriate forms of feedback. A number of visual devices can help in this respect, of which three examples are described below.

Progress indicators

Some mechanisms for controlling human-human dialogue are explicitly aimed at indicating the speaker's limitations so as to modify the other's behaviour. Phrases such as: 'It's on the tip of my tongue' or 'Hang on, I'll get it in a minute' are all designed to suspend dialogue in such a way that it can be restarted shortly and to indicate that the other speaker should wait. Utterances such as 'As I was saying ...' restart the dialogue. Feedback is necessary to avoid inappropriate or unwanted interjections from the other speaker. Analogously, interfaces sometimes appear to 'hang-up' while the computer is carrying out tasks which take time to execute. Without feedback the input devices appear to lock, or worse, the input devices remain unlocked but do not process their inputs until the current activity is finished. This often results in unwanted actions when users repeatedly press keys to get a response.

   A number of devices are used to indicate that an interface is temporarily unavailable or that some activity is in progress. These use methods such as changes in cursor shape, a gradually filling bar to indicate the proportion of the wait time completed, or a count-down clock (see Myers, 1985, for an evaluation of these different methods). First the user is notified that the system is operating (i.e., that it has not crashed); and second, they indicate the length of the delay before the user regains control. Some examples of different cursor shapes that are used to give such feedback are illustrated in Figure 5.1.

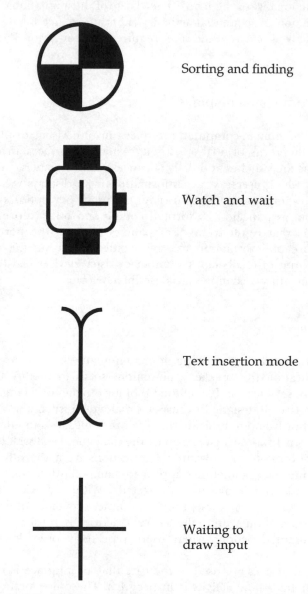

Sorting and finding

Watch and wait

Text insertion mode

Waiting to
draw input

Figure 5.1   Examples of different cursor shapes and their meanings.

Graphical reflection of processes

A good example of how modern interfaces have developed a greater emphasis upon feedback is the use of animation in Graphical User Interfaces (GUIs). For example, the closing of a window in the Apple Macintosh interface is accompanied by a brief sequence in which the window contracts and closes into the icon which represents that file or folder. Functionally speaking, this is an entirely unnecessary elaboration. However, from a psychological point of view, its effect is to provide the user with clear feedback about the process which has just been initiated, both visually and in terms of the time taken. Were that feedback to be absent, and the closing to appear instantaneous on the display, then it is likely that many users (particularly those who had closed the window in error) would be left looking at the display and asking 'What happened?'. This kind of feedback makes it much more likely that users will learn the consequences of their actions and therefore know how to reverse them, and also learn how to avoid closing files by mistake.

Different cursor shapes can be used to demonstrate what state the system is in (drawing mode, selection mode, processing mode, pointing mode, and so on) with little or no cost to the system or disruption of the interface format (e.g., see Smith, Irby, Kimball, Verplank and Harslem, 1982, for examples in the design of the Xerox STAR user interface). These are aimed at reducing errors in which the user misinterprets the mode of the interface and have the advantage that the change of cursor shape coincides with the point of visual attention.

Navigation aids

Many interfaces operate in a number of different states (e.g., the input versus editing modes of certain text editors). In these systems, the same interface is used to present the different modes, and it is necessary for the user to know what state the system is in and how to move between modes. In contrast, a process control panel may be complex, but operators can navigate *themselves* around large banks of displays. Most large process control systems can afford to do this because the operations of a particular plant usually require a dedicated and fixed control room. Self-navigation is easy because we have many more concrete cues to identify where we are and where we have been. However, most computer interfaces have to be flexible enough to support several tasks and are highly limited in the display space available. Therefore they need to have several modes. Moving between modes or states within an interface is an abstract, disembodied concept. It can only be inferred from the feedback given by the system. Feedback is necessary to show where users are in a system, where they can go, and how they might get there.

Early computer systems were particularly poor in providing navigation aids. Some gave no more feedback than '>' or 'ready' to indicate that they were ready to receive command input. Previous transactions were frequently not visible because the screen reset itself or because scrolling text removed them. Hence, the user's had to remember what they had already done, assuming they understood what they had done. Modern systems pay much more attention to this issue, and have much more processing power available to deliver it. For straightforward systems, such as navigating around a single long document, a number of methods are available, including scroll bars, page numbers, and many other imaginative approaches. Navigation around more complex data structures (e.g., menu hierarchies, hypertext and relational databases) presents a greater problem because it is not simply in one direction between the start and end of the space to be searched (Nielsen, 1990). Methods for enhancing navigation through feedback include the provision of graphics to represent the layout of hierarchical filing systems (Spence and Apperley, 1982); 'tagging' specific locations, in which an icon is left on a document in a database and later used to return directly to the document position (Lansdale, 1988); and the use of specific metaphors such as travel, with appropriate graphics to illustrate the metaphor components (Hammond and Allinson, 1989).

Auditory feedback

Sound is a rich medium. It can be manipulated by frequency, timbre, rhythm, volume and timing, among other dimensions. Exploiting sound as a means of extending the diversity and value of feedback in interfaces is an active research topic driven by four main considerations:

Representation of multidimensional data

Multidimensional data, such as the economic parameters associated with a product (volume, profits, orders, turnover, etc.), can be mapped onto multidimensional sounds. Rising pitch might represent greater profits, greater amplitude might represent more orders), and so on. For example, Mezrich, Frysinger and Slivjanovski (1984) mapped economic trend data onto auditory cues such that each economic indicator had a 'voice'. Subsequent dynamic analysis of the trends was aided by the ability to recognize musical patterns. A similar principle has been explored in the use of multidimensional icons dealing with the content, document type and confidentiality of reports (Lansdale, Jones and Jones, 1989). Smith, Bergeron and Grinstein (1990) adopt a similar approach in mixing iconographic representations with acoustic attributes. At the time of writing it would seem that the potential of these approaches has been established, but it remains to be seen which tasks can be supported most effectively by this method.

Interfaces for visually handicapped users

Given the loss of the visual medium, auditory cues have increased significance for visually handicapped users. For example, Edwards (1989) reports a modified mouse-driven interface in which visual feedback is replaced by auditory cues and synthetic speech.

Reducing visual processing by parallel input

A third use of auditory feedback is to spread an information load in a task across more than one input medium. This can be useful in tasks such as air traffic control, where the controller is monitoring radar information in parallel with messages from pilots (e.g., Brown, Newsome and Glinert, 1989). Any method of information display which enhances the distinctiveness of the input channels will increase the user's ability to attend selectively to one or other of the inputs (Broadbent, 1963).

Auditory Icons

Everyday sounds play a large part in our monitoring of ordinary activities. If one throws an empty tin into a trash can, an early dull thud will tell us it is nearly full, whereas a healthy clang will indicate it is empty. In Sonicfinder (Gaver, 1989), many actions are accompanied by auditory icons to add to the feedback. These include a scraping sound for dragging objects, a pouring sound for copying objects (presumably reflecting a gradual movement process), and a whooshing sound for opening files.

Response times

The time taken for a computer system to respond to a user's input is an important source of confirmatory feedback. Consider an application we have developed called COMPUDOC to illustrate question and answer dialogues. The purpose of this application is to extract from patients answers to a number of health-related questions which require a yes or no answer. If we analyse this question-and-answer dialogue, as illustrated in Figure 5.2, we see four points at which a delay can occur. First, on presentation of a question, the

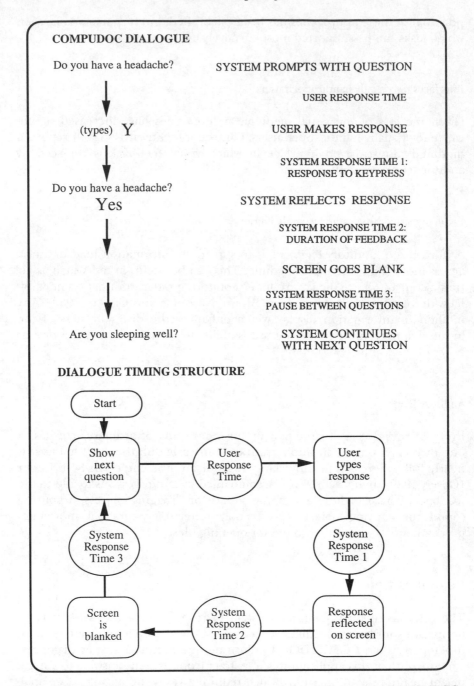

Figure 5.2 The timing structure of a question-and-answer dialogue in COMPUDOC.

user will think before responding. Second, having responded, there may be a finite period of time before the response appears upon the screen (system response time 1). Third, the designer may to leave the question and its response on the screen for a finite period of time (system response time 2). Finally, the designer may insert a delay between clearing the screen and providing a new question (system response time 3).

In an unpublished experiment, users of COMPUDOC were allowed to adjust the three system delays. The preferred settings revealed two common trends. First, all users set the reflection of a simple keypress (system response time 1) to a minimum. This confirms the common finding that users expect simple actions such as keypresses or button presses to have near-immediate feedback (e.g., Shneiderman, 1980). Second, users preferred a delay (typically about two seconds) for the second and third system response time – the time in which the response was displayed and the gap between questions. Some users talked about 'feeling rushed' if one question followed another too quickly. Others complained that if the response delay was too short, they were unable to confirm that the system had accepted the input. In both cases, some level of delay was considered preferable, despite slowing the users down. This example illustrates that response delay is a form of feedback, and serves a number of functions, which are described in Table 5.1.

In our COMPUDOC example, users presented with a question immediately after answering the previous one sometimes reported the feeling of being 'under interrogation', and felt pressured into making fast responses. On the other hand, more confident users were less likely to feel rushed and felt happy to control the pace of the dialogue, by taking as much time as they liked to make their responses.

5.5 Feedback to change users' behaviour

There are two reasons why a system might be designed so as to confront users with unexpected information which is designed to make them take some form of corrective or remedial action to maintain control. The first of these is the need to warn or advise the user of some state of affairs which might require an intervention. These are common in complex systems such as aeroplanes or chemical processing plants. The second reason for negative feedback is that users may attempt operations which are not interpretable by the system or may have unwanted consequences. Feedback in this case might consist of an error message or some dialogue exchange which intercepts the operation before it can be enacted. An example would be a screen message on a word processor such as "The following action will irreversibly delete files. Press OK to continue, or CANCEL to avoid." Users also require feedback to indicate when the system is no longer doing what was expected, even

Table 5.1   Aspects of response times as feedback.

i) Short delays are taken as evidence of cause-and-effect. Generally, as the delay between event A and outcome B increases, the likelihood of seeing them as related is reduced. One effect of delay is to decouple the user's behaviour from its consequences, which increases the difficulties of learning.

ii) Some forms of feedback need a finite time to have their effect. Most menu based interfaces indicate when a menu item is selected – such as a short flashing of the chosen item or other attention-grabbing display that continues for a substantial period of time. The purpose of this is to reinforce that action. If this delay did not exist, users would have no way of checking their input except from the status of the system. Novice users would find the sudden transistions without feedback hard to interpret.

iii) Different actions are accompanied by different expectations of delay. Because users have simple expectations of interface function, they also have simple expectations of how long they will take. Gallaway (1981) produces a range of recommended delays, from 0.5 seconds for simple responses to 15 seconds for executing a complex task. Since these expectations are imported from previous experience, they vary depending upon the user. As technology gradually speeds up, we might also expect the population norms to change also.

iv) Variability in response times is disconcerting. Inconsistency in response delays, such as that encountered by the user of a remote terminal attached to a time-shared mainframe computer, causes errors and annoyance to users (Shneiderman, 1987). This is not simply a matter of being kept waiting for longer, although that is obviously a factor. Teal and Rudnicky (1992) show how users adopt different strategies for coping with consistent response delays when they are consistent which cannot be applied when they are inconsistent. They refer to these as 'automatic', 'pacing' and 'monitoring' strategies: if the delay is negligible, the user can respond immediately; over longer delays, a strategy of matching user responses to complement delay lengths is possible; and if the delays are very long, the user can wait for the delay to be complete. A system that responds inconsistently prevents the user from settling into any one strategy, and the constant monitoring required to check the status of the interface involves a large degree of mental effort.

if it is functioning normally. This is usually because the users have made a mistake which has caused the system to do something different to that intended.

Alarms

In many safety critical systems, such as monitors for intensive medical care, or displays for running a nuclear power station, it is possible to define a set of parameters which must be monitored, and to define certain levels as being abnormal and requiring corrective action or closer monitoring. Alarms, whether auditory messages or a change of state in a visual display, are a method of capturing the user's attention with the ultimate purpose of establishing or reaffirming control of the situation. They therefore interrupt the users actions and provide them with a diagnosis of the problem.

Systems which require alarms are usually complex, with many components and many parameters being monitored. Some of this complexity is added to 'engineer out' the possibility of operator error. Sometimes this leads to what has been described as an 'irony of automation' (Bainbridge, 1987): For much of the time operator's tasks are simple to the point of boredom; but when things go wrong, they do so with bells on, both literally and metaphorically, leaving the operator with a complex and stressful diagnostic task. Many parts of a complex system interact in such a way that an abnormal condition will propagate and cascade throughout the system, activating a large number of alarms. Reason (1990) observes that system failures are often caused by more than one fault, and that some of these part-causes are 'latent' errors, that is, errors whose effect is not immediately apparent.

The use of alarms therefore raises a number of complex issues for interface designers. Amongst these are the need to capture the operator's attention while not distracting them from their jobs; and also the need to provide the operators with detailed information and the consequent problems of interpreting the complicated data. For example, in the Three Mile Island incident, in which a major nuclear meltdown was only narrowly averted, the hundreds of visual and auditory alarms that were activated during the emergency effectively obscured more important sources of information. Indeed, arguably they caused a magnification in the scale of the accident by distracting the operators into actions they should not have taken (Stephens, 1980).

Patterson (1990) has examined the use of auditory alarms on flightdecks and in hospitals. Alarms in these contexts are a means of maintaining control, rather than simply regaining it after a crisis has emerged. An auditory warning should take on the same characteristics as an auditory icon, in which bursts of noise are used to represent a particular alarm state. For example, two long tones which are separated by two short tones, might represent an alarm for 'blood pressure high'. Volume, amplitude and pitch contours can be manipulated to vary the sense of urgency. Greater rapidity with rising pitch and amplitude can sound more urgent, for example, than stable or decreasing pitch and amplitude. Patterson's approach exemplifies several advantages of complex auditory alarms.

First, consideration is given to the acoustic environment in which the alarm is used. Contrary to the 'make it sure, make it loud' philosophy (which often means that an operator's first thoughts are to turn the alarm off), the acoustic properties of the alarm should reflect the task environment. Differences can be seen for example, between an aeroplane flightdeck and a helicopter. In the case of a helicopter, the engine and gearbox noises are intense and auditory alarms need to be distinguishable from the background frequencies, whereas in aeroplanes, the quieter cockpits mean that there is more flexibility possible in the choice of alarm sounds. In a hospital operating theatre, the needs are different: quieter background noises and the need for calm mean that alarms should be softer and less disruptive.

Second, the elaboration of auditory warnings into complex sound sequences removes a good deal of clamour from an alarm state, because the use of silences and rhythms means that less sound energy is used. It also makes different alarms more distinguishable, which is apparently a problem on some existing flight decks. Third, elaborate sound patterns impart meaning as to the nature of the emergency and therefore focus the operator's mind more quickly upon the appropriate action. Finally, auditory icons can reflect different levels of urgency, such as 'general immediate action', 'immediate awareness', and 'information available'. As a consequence, the operator is in a position to *monitor* emerging alarm states as or before they happen. The purpose of alarms can therefore be to maintain control of the interface in changing circumstances rather than to recover control after things have gone wrong. In this sense, as Buxton, Gaver and Bly (1989) observe, Patterson's work converges the development of alarms with the purpose and development of auditory icons as positive feedback.

Forcing functions and error messages

In Chapter 2 we noted that question-and-answer dialogue styles have the advantage of structuring users' actions by limiting their opportunities to go wrong. In fact, a great deal of what we do with equipment and interfaces is structured so that certain opportunities for failure are prevented. For example, one make of coffee grinder incorporates the on/off switch into the protective cover which guarantees that the grounds stay in the grinder. It is natural for the grinder to be used this way (see the concept of affordance discussed in Chapter 4), but the error prevention offered by the design is added value; when error prevention is explicitly designed into a product it is sometimes known as a *forcing function* (Lewis and Norman, 1986; Norman, 1988). Lewis and Norman give examples of the different ways in which forcing functions can operate. These include gagging, in which an inappropriate input is prevented from continuing, as illustrated by this message incorporated into a graphics package:

```
The file you have specified is not a picture file. Please
re-type the filename.
```

An attempt to open this file might otherwise cause a delay or, at worst, a system crash. Other examples of forcing functions are warnings ("The present file has not been saved. Do wish to save before quitting the word-processor?"); and disabling of inappropriate functions (e.g., menu options which are greyed-out when they do not apply). In some direct manipulation interfaces, disabling is accompanied by clear visual feedback. For example, an attempt to copy a file to a locked disc by dragging the file icon to the disc icon might be followed with the icon 'springing back' to its original place, indicating that the action cannot be accomplished.

With the exception of some direct manipulation interfaces, most forcing functions are accompanied by some text explaining what has happened. Some advisory messages indicating that an error has occured can be confusing. For example, the following message is sometimes seen in older MSDOS systems:

```
Seek error reading drive B
Abort, Retry, Fail?
```

This is both a forcing function and an error message. The error occurs if the disc inserted in drive B is of the wrong format or the system is unable to read the disc directory to find the locations of files. Consequently, it would not know where to look for information on the disc. In fact, the message is an invitation to try again (e.g., by replacing the disc) or to stop the action (the difference between Abort and Fail is subtle and need not concern us here). This message is typical of poor error messages in that it fails both to explain what has in fact happened and to indicate satisfactorily what the user should do. Few inexperienced users will understand what a 'seek error' is. The options given for recovery are not self-explanatory, and nor do they indicate that they are activated by keying in their initial letters.

In general, error messages such as this have justifiably received a bad press (e.g., Lewis and Norman, 1986) for being jargonistic, uninformative, hostile in tone, and unhelpful in terms of what the user can do. To be fair, there have been understandable reasons for this problem. In the development of interfaces, the need to produce a working system is imperative. Equally, the pressure to hold a system back from the marketplace simply because the error messages have not been carefully checked is not strong. Also in older technologies, the time and screen space taken in possibly lengthy explanations of an error might be seen as counter-productive; especially when one considers that many older interfaces were designed with trained users in mind who might be expected to cope with such messages. In all, the production of useful messages (or any at all) was of low priority.

Modern guidelines tend to emphasise the need to give clear and informative messages (e.g., Shneiderman, 1987). The requirement for better error messages has been prompted by two main factors. First, the population of interface users has expanded enormously, with fewer users being 'expert'. The expansion of word-processors in particular has resulted in a large user population (upon whom the manufacturers rely increasingly for sales) who are not formally trained. Such discretionary users are less likely to develop their skills if learning is inhibited by obstacles such as impenetrable error messages. There is therefore greater incentive for manufacturers to be as careful about the design of error messages (as all other aspects of interfaces). Second, there has been a realization that human beings are highly prone to error which is almost impossible to eradicate, even with training. The result is a greater interest in the understanding of different types of error and their causes, accompanied by a need to develop ways of mitigating their effects.

5.6   Repair: Regaining control and dealing with errors

Humans as natural error-makers

The tendency to make mistakes is a good example of the compromises that have to be made in human performance. On one hand, it is advantageous to be as quick in thought and deed as possible. On the other hand, the quicker we think, the more likely we are to make errors. This is often known as a *speed-accuracy trade-off*, a relationship which is illustrated in Figure 5.3. This

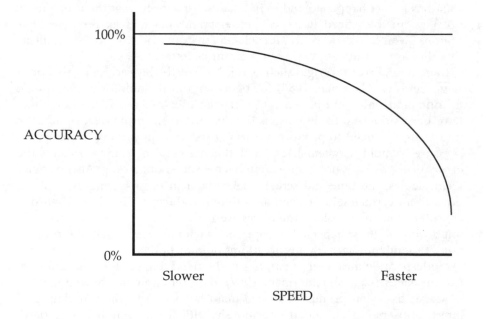

Figure 5.3 An idealized relationship between accuracy and speed in human performance.

graph reflects the performance in experiments of subjects who are asked to make decisions or judgements as quickly as possible, and shows how the speed of response is related to the accuracy of response. The basic outcome of such experiments (e.g., Pachella, 1974) is that, the faster humans are asked to respond, the less accurate their performance. Conversely, the more accurate they are asked to be, the longer are their response times. An attempt to eradicate errors can lead to unacceptably long response times.

In practice some errors appear to be unavoidable, and incentives to avoid errors merely lead to increasingly long response times. It is also worth noting that most of the errors we commit in life are not serious. Even in the area of driving, where one would imagine the incentives to avoid error are high, most errors are of no serious consequence. In general, the attempt to maintain totally error-free performance is usually not cost-effective, and circumstances where it is required can be stressful. Most human life involves learning to tolerate, rather than avoid, errors. The message for designers is that they should structure interfaces in such a way that some errors should be tolerated rather than avoided, and should be repairable when they occur. Over-emphasis on error-avoidance is unpleasant for the user and is likely to lead to poor learning strategies.

In order to know how to design interfaces which tolerate user error, it is necessary to identify the types of error which might be expected (e.g., Norman, 1981b; Reason and Mycielska, 1982). There is a broad consensus about the classification of errors. The main distinction is between *mistakes* and *slips*. Mistakes occur when the user makes a wrong decision about what action to undertake which then leads to an error. Slips occur when the user makes a correct decision about what action to undertake, but then undertakes an erroneous action instead of the intended action.

Mistakes

Mistakes are harder to define than slips, since it is the intention rather than the execution of an action which is said to be at fault. As a result, often users do not recognize their mistakes until some time after they occur, since the actions they undertake are consistent with their intentions. Some mistakes arise simply from ignorance of a task. For example, novice users may not realize the need for saving new document files and therefore may lose their work. Avoiding mistakes often involves training, such as learning to make regular backups of material.

Many interfaces are designed to prevent mistakes by limiting the opportunities for the user to issue an erroneous command. Structured dialogues using forms or question-and-answer styles reflect this limitation, and forcing functions (see above) also serve the same purpose. To some extent, mistakes can be minimized by these methods. However, complex interfaces that require user expertise, and which allow flexibility of use, will also create the potential for mistakes to occur. In circumstances where safety (or for that matter, a large sum of money) is at stake, efforts to minimize the potential for mistakes might necessitate the design of positively 'user-hostile' aspects to an interface, requiring users to conform to cumbersome or involved procedures, possibly also requiring actions to be performed in parallel with other users.

Although we can attempt to minimize mistakes by training or by designing alarms and override mechanisms, as often as not organizations also apply sanctions to enforce stack operating regulations. However, nuclear disasters such as Three Mile Island and Chernobyl, as well as accidents such as at Tenerife Airport in 1986 (where a misunderstanding between air traffic control and pilots led to a collision of two 747 jets on the runway with the loss of hundreds of lives) are proof that, despite training, design and legislation, one can never totally eradicate mistakes. Analyses of disasters and near-disasters are beginning to establish the means to identify *post hoc* the causes of mistakes (e.g., Reason, 1990). Psychologists are also developing methods to support the design of safety-critical systems. However, the design of mistake-free systems probably cannot be achieved in the foreseeable future.

Slips

Slips are the most common source of error, for both novice and expert users. They can occur for a number of reasons that have in common a failure to achieve what was intended. Norman (1988) lists six reasons, of which we describe three below in detail (see also Rouse and Rouse, 1983, and Reason, 1990). Understanding why slips happen gives us a better insight into how they can be avoided or repaired.

Capture errors

The best known example in the psychological literature of a capture error is offered by James (1890), who reported going upstairs to change for dinner, washing, and then going to bed, leaving his dinner guests waiting downstairs. The initial sequence of going upstairs and undressing 'captured' the behaviour of going to bed. Generally, capture errors occur when two or more well-learned sequences share common elements and through some lapse of concentration allow the initial intention to be taken over by the wrong action. Green, Payne, Gilmore and Mepham (1984) describe a computer-based example in which a command W writes to a file and a command Q quits the word processor. The concatenated command WQ allows the user to backup and then quit, but the drawback is that as skill becomes automatic, the user wishing merely to back-up captures the command WQ and finds that the application has quit.

Description errors

Some slips appear to be failures to specify an intention unambiguously. For example, one might distractedly pour the coffee jug rather than the milk jug over the cereals. This is a description error because the absent-minded mental description of an object ('something to pour out of') fails to be sufficiently precise to distinguish between two alternatives. Like most slips, description errors are more likely to occur when a person is tired or distracted and can be seen as a lapse of attention. In interfaces, this type of error more often occurs when two objects (such as filenames, switches, or icons) are similar in some way. Description errors are therefore less likely when distinctiveness is maximized, hence the addition of coding mechanisms such as shapes, shading and location to enhance distinctions between objects. This is illustrated by the use of icons in Figure 3.10.

Mode errors

Mode errors occur when the interface or object being handled responds to user actions in a different way while maintaining the same external appearance. For example, some older word processors have two modes of text input. In the INSERT mode, any typed characters are inserted at the chosen point, and the surrounding text parts to take the extra characters; in the REPLACE mode, the same input has the effect of overwriting and replacing existing text. Not surprisingly, errors in which text is unwittingly deleted or left intact when not wanted are common.

Mode errors often lead to a breakdown in human-computer interaction which is similar to the silence that occurs between two people both waiting for the other to speak first. This happens when the user thinks a system is in 'doing' mode when it is, in fact, waiting for input. For example, our FAX machine has a very small display which is singularly uninformative about the state of the machine. Because delays in FAXes are common, one may wait for a very long time before remembering that the FAX machine actually requires the user to press the START button before proceeding.

The origin of such errors is a failure to keep in mind (or simply never knowing) which mode the system is in. This is sometimes caused by a failure of the system to indicate what mode it is in. This can be rectified by increasing the cues which differentiate between the different modes, such as colour coding, screen formats, and the shape of cursors. These act as obvious reminders of the status of the interface. Sometimes it is not possible to add cues to different modes, for example Norman, 1988, describes a button on a wrist-watch which serves two purposes: in the 'stop-watch' mode it resets the stop-watch to zero; in other modes, it illuminates the display. The resulting error is predictable: having run home in record time in the dark, the user looks at the watch to find it reset. The problem is that the wrist-watch is packed with much more functionality than its display and input/output devices can handle: a modal system is inevitable because the buttons have to serve more than one purpose.

This example illustrates how technological or physical limitations can lead to design options that increase the likelihood of errors. The quality and resolution of displays has developed alongside an increasing flexibility of input devices. As a result, the pressure to design modal systems (except in circumstances like wrist watches, where the physical constraints of size are paramount) has decreased. It is no longer necessary to blank-out a display to operate in another mode, because an additional overlaid window is often a suitable alternative, leaving the previous modes evident and accessible in other windows. Equally, few word processors now use the insert/replace modes because the fast display processing means that inserted text can appear to move all other text to the right without difficulty. Also, the change of dialogue styles that has accompanied use of the mouse as an input device has made these modes redundant: the effort of specifying areas of text to be deleted is low in mouse-driven interfaces compared with key-command systems.

Other slips

Norman describes three other categories of slip which we summarize briefly since they are less likely to occur in interface use: *data-driven errors* (in which semi-automatic responses to data cause an unintended response, such as attempting to ring someone by their room number rather than telephone number); *associative activation errors* (including 'Freudian Slips' in which something one is supposed not to say is precisely what slips out); and *loss of activation errors* (when we forget what we were going to do).

Many slips are also simply mechanical failures in which the wrong key was pressed, the mouse button released too early, or the cursor not placed in quite the right place. These are distinct from the errors/types classified by Norman in that they are not psychological in origin.

The value of the UNDO command

Interfaces should be designed to minimize the likelihood of slips and, where slips do occur, their effects should be easily reversed. Many systems in the past have presented precisely the opposite profile, and have fairly earned a reputation for being user-hostile. For example, we discussed in Chapter 1 a case study by Lansdale and Newman (1991) of an office and warehouse management system in which serious errors were really rather easy to make. The system operated in different modes, with different operations having the same command input, depending upon the state of the system. It is not hard to see how slips could occur in this case. Further, some of the slips were very expensive – a mistaken activation of the basic ledger program locked the system to all other users for several hours and was not easily stopped. It is important to minimize the seriousness of slips because they are something that users cannot avoid, and the less time spent correcting them, the better.

Another reason why we should concentrate upon avoiding rather than remedying slips is because of their relationship with levels of *arousal*. The term arousal is used by psychologists to refer to the amount of anxiety, vigilance, concentration and other aspects of mental activity that are evident in peoples' behaviour. Studies of learning have consistently shown that the relationship between effectiveness of learning and arousal is an inverted-U shape: too much or too little arousal results in poorer learning (e.g., see Fisher, 1986; Eysenck, 1977). This is illustrated in Figure 5.4. If a task is

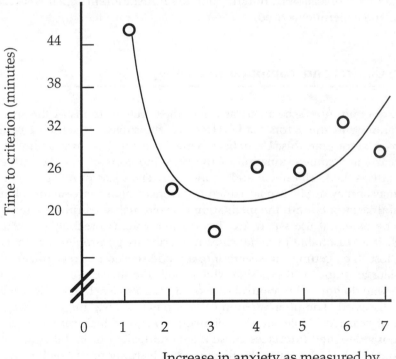

Figure 5.4   (After Matarazzo *et al.* 1955, cited in Fisher, 1986) An illustration of efficiency of learning as a function of anxiety, showing a U-shaped relationship between anxiety and learning with faster learning at intermediate levels of anxiety.

boring, uninteresting, or too easy, or if it is too difficult and the source of anxiety, then we learn less than if the task was interesting and mildly challenging. Clearly, if the cost of errors is great, then anxiety levels rise and learning is inhibited. Interfaces which exact a penalty for error are therefore harder to learn.

Another finding is that highly active subjects who explore an environment and take risks with it, tend to learn more then passive subjects (e.g., Carroll, 1982b). If errors make the price of exploring too high, then users learn less quickly. Maintaining the user's control over an interface and allowing them to explore it by reducing the cost of errors is therefore an important objective of design.

The introduction of the UNDO command in modern interfaces provides a feature which allows users to take risks in exploring a system safe in the knowledge that the results of any action can be undone if they turn out to be undesirable (for a review, see Yang, 1992). The UNDO command restores a system interface to its former state after an operation has been carried out. For the user, this allows a different mode of operation in which risks can be taken and experiments tried.

5.7   Control and 'computer phobia'

Control of an interface can be assessed objectively in terms of the speed of operations and the errors made. However, the subjective sense of being *in control* is more than this. Users have perceptions of their own ability to execute tasks and some estimation of the risks and costs of mistakes, which in turn affects their self-esteem and confidence. The sense of being out of control may or may not be accompanied by errors or poor performance, but it is certainly associated with the inclination to avoid such stressful situations.

Some potential users may avoid computer systems precisely to avert this stress. It is reasonable in many cases to ascribe usage problems such as getting 'lost' (i.e., getting the system into a mode you do not recognize) or losing data to 'traps' in the design which catch the unwary or inexperienced. Users who do not feel in control of a system are also less likely to learn, as we have discussed. The phenomenon of 'computer phobia' might be seen as a form of 'learned helplessness' (Seligman, 1975), in which early experiences of poorly designed interfaces cause a general feeling of inability to control computers which decreases the likelihood of learning to use such systems in the future.

Social factors also play a role in holding uncertain users back from using computer systems. A characteristic of the introduction of information technology has been the way in which technology is seen as the province of the expert or 'boffin'. For many older people in particular, computers were seen as exclusively used only by qualified individuals. Additionally, many of these 'expert' users have actively exploited the technology as a vehicle for organizational status and power. Indeed, some organizational behaviour seems to suggest a tacit conspiracy among the computer users to ensure that this remains true (Eason, 1988). Unfortunately, errors are often highly visible in the sense that the user inevitably has to recognize failure publicly by seeking help. Such factors increase the difficulties of learning and put a greater premium on the design of interfaces to allow for control at all levels of expertise.

5.8   Conclusions

This chapter has examined three factors which contribute to the maintenance of dialogue: prediction, feedback and repair. Prediction is dependent on two main factors: the consistency of interface design, and the understanding that a user constructs of the interface. Feedback is necessary to confirm that a users actions are executed as intended and also to interrupt or prevent inappropriate actions. This feedback should be multidimensional and should not be minimized in the mistaken belief that it increases the information load to be handled by the user. Well designed systems provide substantial feedback without much of it being overtly intrusive. When control breaks down, it is necessary to repair dialogues. To do this requires an understanding of the types of error which might occur, and the provision of mechanisms such as the UNDO command which enable users to recover control with ease.

Control is not synonymous with expertise. While experts know more about a system and can do more tasks more flexibly, it is also possible for them to experience a lack of control. This is particularly likely if the feedback is minimal or the user's expertise is such that many of their operations are carried out automatically. Control is therefore something designed into the system, and is not simply something to be learned.

Part 3
Understanding User Skills

In this part of the book we concentrate upon human skills and the way in which they are applied to the use of interfaces. Skill is a complex concept which refers to a wide range of human activities, and we approach the issues involved from three perspectives. Chapter 6 considers skill in terms of the learning of procedures – the development of sequences of activities which enable experts to carry out complex tasks rapidly and with minimum mental effort. Chapter 7 looks at skill in terms of analysis and understanding. This view of skill describes how we think our way through complex problems, and how we use what we know to do it. Chapter 8 discusses the role of exploration as an element of human skill, and how this is supported in the innovative design of interfaces.

CHAPTER 6

Skill as procedures

Overview

This chapter introduces the topic of user skills, and examines the extent to which interface skills are composed of procedures, that is, sequences of actions and decisions learnt by users for specific tasks. We describe how procedures become automated with increasing expertise, a process which lowers the mental load faced by a user performing a familiar task. The disadvantages of automation are also described, notably the fact that task performance becomes increasingly inflexible, task-specific and difficult to interrupt or correct. The acquisition of procedural skills seems to involve a process of learning-by-doing. To examine the learning of procedures, we use Anderson's (1983) ACT* theory as a framework for describing the stages of skill acquisition. Limitations of this framework are described, some of which are specific to ACT*, others which are general to a view of skill as consisting solely of rote-learned procedures. Finally, we discuss methods for training procedural skills, focussing on the roles of feedback and part-task training, and introducing the minimal-manual approach of Carroll (1990) to developing documentation.

6.1 Introduction: what are user skills?

Skill comes in many different guises: from that needed to perform simple tasks such as recalling the required command name for a text editing operation; through skills for carrying out a series of actions to shut down a pumping operation from a control panel; to diagnostic skills used in the complex tasks of understanding, debugging and testing computer programs. A number of types and levels of skill may be exhibited by different users of an interface. For example, copy typists are highly skilled in their use of the keyboard, as measured by the speed and accuracy with which they can input text. However, keyboard skills are arguably less significant for an academic author writing a journal article, for whom skill at planning documents using an outliner would probably be more valuable than sheer typing speed. Thus a user of a word processor who appears highly skilled at tasks that involve significant quantities of typing will not necessarily be skilled at tasks which involve document structuring, and vice versa.

The difference between speed typing and planning using an outliner serves to illustrate a distinction often made in the literature (e.g., Payne, 1988) between skill-as-procedures (discussed in this chapter) and skill-as-understanding (discussed in Chapter 7). Some skills involve the reproduction of rehearsed procedures to carry out known tasks, whilst others involve the creation of new sequences of actions based on the users' understanding of the interface. Both these views suggest that skill comprises the application of knowledge about interfaces that users acquire and hold in their memories.

A third type of skill does not reside solely in the memory of the user, but instead reflects the users' ability to extract information from the interface itself. For example, in Chapter 4 we described the study by Mayes *et al.* (1988) that showed how experts had a poor recall of command menus which they nonetheless were able to use easily. Howes and Payne (1990) suggest that, rather than recall the name and position of the required command, experts indulge in 'menu flicking', which involves a rapid search through all the menus presented on screen for a menu item that matches their current goal. The 'skill' exhibited by experts appears to be their ability to extract the required information at the right time from the menu display. In other words, skill described here involves the *exploration* of an interface, an aspect of skill discussed in Chapter 8.

The way in which the three aspects of user skill are discussed in the chapters of Part 3 is outlined in Figure 6.1. It is often the case that psychological theories treat procedural, understanding and exploratory skills as reflecting distinct classes of knowledge. For example, a distinction is often made between *procedural knowledge*, that is knowledge about *how* to carry out a task, and *declarative knowledge*, that is knowledge about *what* constitutes the task and the environment in which it is carried out. Some theories (e.g., Anderson, 1983) go further, suggesting that the distinction between declarative and procedural knowledge also defines the difference between novice and expert knowledge. As well as procedural and declarative knowledge, psychological theories often describe the existence of *strategic knowledge* (e.g., Newell and Simon, 1972). Strategies are essentially general methods (as opposed to the highly task-specific methods of procedural knowledge) which provide useful mechanisms for carrying out a range of tasks. For example, menu-flicking is a strategy for finding commands which is based on the user's assumption that it is more cost-effective to search a display for required items than to learn their positions and names.

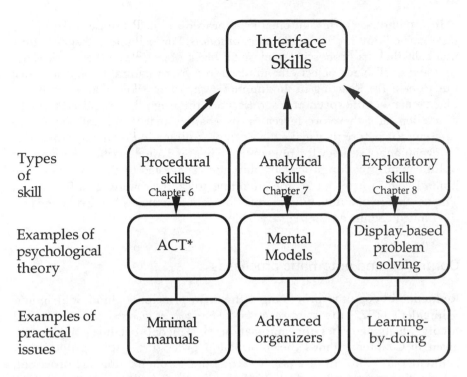

Figure 6.1   An overivew of Part 3, illustrating how each view of skill is associated with a different theoretical approach and set of practical issues.

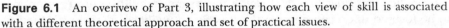

6.2   Expertise as automated procedures

Often when an experienced operative performs a complex task, such as typing a letter, driving a car or using a knitting machine, they seem to do so in a way that is both fast and error-free and yet requires little or no mental effort. This is because they employ frequently rehearsed procedures to carry out routine tasks. Driving a car is a good example of this type of skill. When one first learns to drive, the process of changing into first gear and setting off seems to be a difficult procedure involving a number of sequenced steps. For an expert it is so simple a task that they just do it without having to think of individual steps, like looking in the mirror, depressing the clutch pedal, moving the gear stick, gradually releasing the clutch, and so on. Skills like this appear to be *automated*, that is, to be performed without effort or conscious control. Repeated practice of many interface tasks, such as logging on and off a bulletin board, text-editing actions, or selecting commands from menus, leads to the apparent automation of the skills necessary to perform the tasks.

In evolutionary terms, automating a procedure makes sense for frequent tasks carried out in unchanging environments. For example, a frog will automatically flick its proboscis at any small black object, since the chances are that any small black object encountered in a frog's natural habitat will turn out to be a fly. Reacting in an automatic way to anything that is small and black enhances the speed of response since the frog is not required to make a decision, and therefore increases the chances that a frog will be able to catch insects before they make their escape. Automation is equally sensible as an evolutionary strategy for humans. In Chapter 1 we described the concept of human information processing, and it was suggested that people they have limited cognitive resources for attending to, and processing, information. The advantage of automating a procedure is that it makes fewer demands on the limited cognitive resources available.

Controlled and automatic processes

Reduction in cognitive processing is the main principle behind Shiffrin and Schneider's (1977) theory of *controlled and automatic processes*, a psychological theory which describes the automation of skilled procedures. Figure 1.2 given in Chapter 1 shows the information processing system proposed by Schneider and Shiffrin. If a task is carried out through automatic processes, information can flow directly through the system without requiring attention to be focused upon carrying out the task. In Schneider and Shiffrin's theory, automatic processes do not require attention and are not limited in terms of the capacity for carrying them out, but once automated are not modifiable. Controlled processes, on the other hand, require attentional resources and are of limited capacity, but can be used in different contexts.

To demonstrate controlled and automatic processes, Schneider and Shiffrin carried out an experiment using a letter discrimination task which is described further in Figure 6.2. To summarize their results, they found that subjects were able to discriminate a letter from a group of numbers significantly faster than they could discriminate a letter taken from the first half of the alphabet from a group of letters taken from the second half of the alphabet. Discriminating letters from numbers is, according to Schneider and Shiffrin, an automatic process which is practiced thousands of times in the course of everyday life. Therefore, subjects can look at the whole display of numbers in a single glance to see if a letter is amongst them. However, it is rare to have to discriminate letters from each half of the alphabet, since they tend to be grouped together in words. Therefore, discriminating letters from different halves of the alphabet is a controlled process, which requires conscious attention to each of the letters given in a display.

The finding that it is easier to recognize a letter amongst numbers than a letter amongst other letters is not in itself of particular significance.

Task: Subjects were shown a set of cards one by one in rapid succession, and had to identify whether a target letter was amongst the figures on any of the cards in the set. Stimulus size was varied by presenting cards with either 1, 2, 3 or 4 figures. Stimulus complexity was varied by requiring subjects to identify, either a target letter from the second half of the alphabet appearing amongst distractors taken from the first half of the alphabet, or a target letter appearing amongst distractors taken from the numbers from 0 to 9. Examples of stimuli (with four figures per card and target letter 's') are as follows;

Letter - Letter stimuli: Letter - Number stimuli:

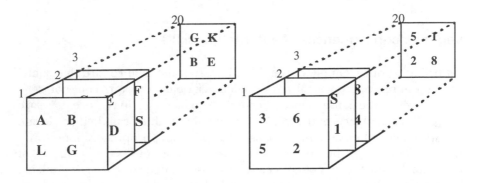

The speed of card presentation was gradually decreased to the point where subjects could reliably identify whether or not the target letter was present. Schneider and Shiffrin found that they could present cards containing the letter-number stimuli as fast as 80 milliseconds per card, regardless of the number of figures on each card, and still get reliable identification of the target presence. Whilst the presentation rates for cards containing letter-letter stimuli with 1, 2, or 3 stimuli were comparable with letter-number stimuli times, the maximum speed for presenting cards containing letter-letter stimuli with 4 figures per card was 400 milliseconds. They argue that letter-number discrimination is an automatic process which is unaffected by the size of the stimulus, because the figures can be scanned in parallel. Letter-letter discrimination, on the other hand, is a controlled process, and is slowed when the stimulus size is large because each figure must be scanned in series.

Figure 6.2   The letter discrimination task used by Schneider and Shiffrin (1977).

However, in a second experiment, Schneider and Shiffrin took the difficult letter-letter task from their first experiment, and made their subjects repeat the task more than 2000 times. They found that repeated practice led to performance which was as good as in the letter-number discrimination task. In other words, repeated practice was shown to reduce the attentional demands of the task by allowing previously controlled processes to become automated.

The development of expert skills for carrying out interface procedures (e.g., command selection) involves the automation of actions and decisions through repeated practice, leaving the expert free to concentrate their attention on other aspects of the task in hand. A consequence of automation is that the ability to carry out a number of complex tasks at one time is often enhanced by existing knowledge of practiced procedures. For example, skilled typists can type a piece of text whilst repeating back another verbal message simultaneously (Allport, Antonis and Reynolds, 1972) and pianists are capable of carrying out the complex task of sight-reading a musical score whilst conducting a conversation (Shaffer, 1975).

Negative consequences of automating skills

Returning to our example of the fly-eating frog, the chances are that sometimes it will react automatically to a small black object that turns out not to be a fly. However, the consequences are unlikely to be too serious, since there are few small black objects in the frog's natural habitat which can do it any harm. Unfortunately, when automated skills are applied inappropriately in performing interface tasks, the consequences are often serious. One of the most important characteristics of automatic processes is that they are not available for conscious inspection. If you don't think about a procedure as you perform it, then you cannot monitor its performance. The disconcerting experience that drivers sometimes have, of being completely unaware of the journey they have just completed when following a familiar route, is an example of an absence of monitoring. Of course, the advantage of this is that one can think of other things. However, when a procedure is not monitored, it becomes less flexible and more prone to the sorts of errors discussed in Chapter 5.

If a familiar (i.e., automated) task is altered in some way such that the old procedures do not apply, then poor performance sometimes results. Consider, for example, the use of on-board telephone systems in vehicles. The problems of holding a receiver whilst driving have led to the development of hands-free car telephones. Whilst these seem to remove some of the ergonomic problems of using telephones whilst driving, there remains a major psychological problem with the use of any telephone whilst driving. Under normal conditions, a driver is able to cope well with the demands of driving whilst conducting a telephone conversation, because most aspects of driving skill are automated for experienced drivers. However, when an unusual situation occurs, such as a burst tyre or a child running into the road, attending to the phone message suddenly becomes a major burden which may reduce the speed at which the driver will respond to the unusual driving event. The simple ergonomic fix of a hands-free receiver does not fully address the real problem of vehicle telephones, which is the need to switch attention from the conversation to the road during emergencies.

The lack of flexibility of automated procedures was demonstrated by Schneider and Shiffrin (1977) using their letter discrimination task. Having trained subjects to spot a letter taken from A-M in a display of groups of the letters N-Z, they then reversed the task, requiring subjects to spot a letter taken from N-Z in a display of groups of the letters A-M. Performance was far worse with the reversed stimuli than it was before subjects had practiced the previous task. In fact it took over 1000 trials to get back to the level of performance subjects had at the start of the experiment. In other words, the effect of practicing the recognition of a letter taken from A-M in a display of groups of the letters N-Z was to impair the reverse task relative to no task practice.

An example of an interface where similar problems can arise is the numeric keypad for calculators. Two possible layouts are illustrated in Figure 6.3. A skilled data entry operator who has been trained using one keyboard will perform worse on the other keypad, in terms of speed and errors made, than someone who has not regularly used a numeric keypad. This is an example of *negative transfer*, in which the application of skills learned in one situation actively impairs learning and performance of the same task in a similar situation. Negative transfer is related to the concept of *interference* (Postman and Underwood, 1973), in which information that is memorized in one task interferes with the recall or recognition of information presented in another task.

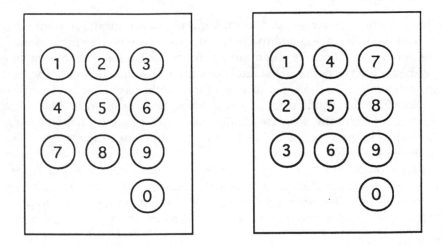

Figure 6.3 Alternative numeric keyboard layouts. Skilled operators switching from one to the other will tend to make more errors than inexperienced users because training with one keyboard becomes automated and therefore will not transfer to the other keyboard.

6.3   How are skills acquired?

When a person encounters a new interface, they may have an intended outcome in mind but a limited knowledge of how to use the interface. For example, consider a person who has never used a word processor before, and who
is required to learn to use one as part of their job or education. The chances
are that they may know little about computers, though they may be familiar
with keyboards, having used a typewriter in the past. Three ways that they can
find out how to use the word processor are as follows:

Reading instruction manuals

Reading the instruction manual seems a reasonable idea in principle, but in
practice few people read manuals from cover to cover before starting to use a
new interface, and it doesn't always help if one does. This is particularly the
case when the user of the word processor wishes only to carry out a simple
task in the first instance, like inputting, printing and saving un-formatted
text. Despite improvements made over the past few years in the documentation that comes with commercial word processors, manuals are still frequently cumbersome, poorly laid out and inadequately explained. As a
consequence, people often turn to the manual only when they are stuck or
have made an error from which they cannot see how to recover.

Experimenting with the system

Perhaps the most common and most natural way of acquiring interface skills
is to turn the machine on and try things. It satisfies the user's inquisitiveness
and desire for rapid results with minimal effort. In Chapter 8 we will see how
hypermedia interfaces are deliberately designed to support exploration as a
learning strategy. However, the success of this method relies on an interface
being designed in such a way that it is apparent how it should be used (that
is, it should exhibit the property of affordance, described in Chapters 4 and
5), as well as offering extensive feedback on the outcome of the user's activities. Additionally, the user may find it useful to have at least some prior
knowledge of the interface. However, prior knowledge can be applied inappropriately. For example, as we shall see in Chapter 7, a novice user very
rapidly runs into difficulties if they base their understanding of how to use a
word processor entirely on their experience with typewriters.

Receiving guidance from experienced users

An effective source of information for novices learning how to use a word
processor is to ask someone who already knows how to use the machine.

Having a guide who knows the system sitting next to a novice whilst they learn to interact with an interface is an immediate source of instruction, feedback, wisdom and encouragement that could never be matched by any manual. Of course, this relies on having the right person available at the right time. Surprisingly, an expert is not always the best person to act as a guide, since experts may choose the wrong level of detail in instructing novices. Also, as we shall see later in this chapter, experts are not always able to verbalize the knowledge that underlies their expertise. Guided tuition also relies on the novice being able to understand and then recall at the appropriate time the pearls of wisdom imparted by their guide.

In practice, a novice will acquire skills by a combination of these methods. However, with all three approaches the learner may acquire incomplete or even faulty knowledge which will impair their ability to learn new tasks. It is the task of both interface and training method designers to enable successful use of a system whilst minimizing the likelihood that the learner will acquire incomplete knowledge or misconceptions. It is unfortunate that in many software development projects, interface design and training design are treated as separate stages in the development of a new system.

It is widely recognized that certain skills must be practiced if they are to be learned. For example, you cannot learn to ride a bicycle by reading a book about it; you have to go through the sometimes painful process of attempting the task, perhaps helped by advice from someone who already knows how to do it and who can offer advice (e.g., that one should be moving in order to retain one's balance). The same is true with many interface skills, such as learning to write computer programs, editing text with a word processor or drawing pictures with a graphics package. In general, *learning-by-doing*, as this form of skill acquisition has come to be known (after Anzai and Simon, 1979), requires a certain amount of instruction or understanding about the skill. Nevertheless, the emphasis is on people carrying out tasks in order to refine methods for performing them.

The ACT* account of skill acquisition

A psychological framework which describes how skills are acquired though learning-by-doing is Anderson's (1983) ACT* (Adaptive Control of Thought). The framework has been developed through empirical research into interface skills, notably those of computer programming and text editing, although the principles of the framework are supposed to generalize to all cognitive skills. The framework is ambitious in attempting to provide an account of how all cognitive processing is carried out, and as such is open to criticisms about its testability. Nor is it unique in many of its principles (indeed, it contains many of the concepts about human performance developed by Newell, 1973). For the purposes of this discussion, however, it provides a useful description of many of the phenomena associated with learning-by-doing; notably how people move from novice-like behaviour to a state of expertise through the performance and practice of procedures.

ACT* describes human knowledge in terms of *production rules*. A production rule consists of two components - a description of one or more conditions and a set of actions that are associated with these conditions. The conditions describe a goal of the problem solver, whilst the actions can either be sub-goals, methods or knowledge that accomplish the goal. For example, the following production rule might be used to make analogies between new problems and previous examples:

IF   the goal is to write a solution to a problem and there is an
         example of a solution to a similar problem
THEN   set as subgoals
         1. to compare the example problem to the current problem
         2. to map the example solution onto the current problem

If the conditions of a production rule are satisfied by matching them with the goals of the current task, then the actions entailed by the rule are undertaken. The successful activation of the rule described above creates a further sub-goal to compare the example problem to the current problem, which will then be matched against another general production rule. Once this has happened the second subgoal, to map the example solution onto the current problem, can be undertaken directly without calling on further production rules.

ACT* describes the acquisition of expertise as the development of production rules for carrying out specific actions which accomplish specific tasks. These rules are then activated whenever the current task has goals which match the conditions of the rules. Anderson (1983) proposes three stages in acquiring a cognitive skill; a declarative stage; a stage of knowledge-compilation; and a procedural stage.

The declarative stage

In the declarative stage of learning, novices apply general production rules to the information that is given about a task. General production rules are methods which can operate in any context for carrying out processes such as reasoning-by-analogy or trial-and-error. The production rule shown above, for making analogies between new problems and previous examples, is one of the most important general production rules that are used by novices in acquiring a new skill. Although novices must use general rules to undertake new tasks, they place heavy demands on short-term memory since they require the processing of large amounts of information (notably in the form of related examples). As a consequence, performance is slow and tends to be error-prone because novices forget important details when memory becomes overloaded.

In order to investigate the declarative stage, Anderson, Farrell and Sauers (1984) studied the acquisition of LISP programming skills. LISP is a language used in Artificial Intelligence programming. It is specialized for manipulating lists of symbols (words, letters, etc.) and is therefore particularly suited to developing applications for natural language understanding. Anderson *et al.* describe in detail the efforts of one novice to tackle a new LISP programming problem, which was to write a function that returned the first element in a list. Before tackling the problem, she received instruction about some of the commands for defining functions which are used in LISP programs, notably DEFUN, which is used to indicate that a new function is being defined, CAR which returns the head of a list, and CDR which returns the tail of a list. Examples of the application of these functions are as follows:

CAR(a b c) = a

CDR(a b c) = (b c)

and   DEFUN(FIRST(LIST1))
        CAR(LIST1)).

The last example illustrates the solution that the novice finally generated for the problem, which roughly states "Definition of a new function called **FIRST** which operates on the parameter LIST1; when evaluated, this function returns the first element of the list represented by the parameter LIST1."

The novice also received some examples of LISP programs, for converting Fahrenheit to Centigrade and for exchanging the position of two items in a list, and a template for writing LISP functions. The three types of knowledge sources used by this novice (language keywords, examples, and the template) represent the declarative knowledge that is typically given to novices in textbooks and instruction manuals. The way in which the novice tackled this problem is illustrated in Figure 6.4. The boxes indicate the goals that were set by the novice at each stage, the arrows indicate the application of a production rule to try and achieve each goal, and the ticks and crosses indicate whether the application of a rule was successful or not.

The novice attempted to solve the problem by mapping the template and the example solutions onto the current problem. When errors occured, she either sought another example and tried to map that onto the current problem, or she recalled a similar error that she had made in the past and tried the same fix that was used to get over that error. She eventually achieved a correct solution after a number of attempts. Anderson *et al.* argue that the solution was achieved principally through the application of the general production rule for making analogies. Perhaps the most significant and surprising observation was that *understanding* the current task appeared to play little or no part in her attempts at a solution. Instead, she relied almost exclusively on making superficial analogies and other behaviours which might best be described as trial and error.

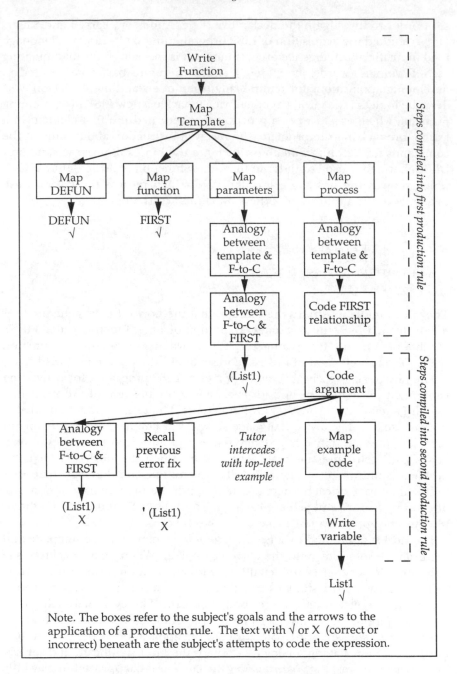

Figure 6.4 The steps taken by a novice in writing a LISP expression to return the first element of a list (after Anderson, Farrell and Sauers, 1984).

The knowledge-compilation stage

The second stage in Anderson's (1983) framework is one of knowledge-compilation, in which general rules are converted into task-specific rules. Task-specific rules make fewer demands on memory because declarative knowledge is 'compiled' into the rules. In other words, instead of relying on general procedures to process the information that is relevant to a particular problem, the novice acquires specific procedures which already contain relevant information for solving the problem. Knowledge-compilation relies on feedback about the success of each rule application. When a general rule is selected for application, its actions either achieve the desired result or it fails. If it succeeds, then a task-specific rule is created that essentially contains a description of the exact context and method in which the general rule was applied successfully. If it fails then another general rule is chosen for application.

Knowledge-compilation consists of two main processes: *composition*, in which a number of rules are combined to make a single rule; and *proceduralization*, in which task-specific information is added to a rule. Essentially, the rules become customized to perform a sequence of steps necessary to carry out the task through the processes of composition and proceduralization. For example, Anderson *et al.* suggest that, in solving the programming problem of writing a function to return the first element in a list, their novice compiled the following task-specific production rules:

> IF   the goal is to write a function of one parameter
> THEN   write *(DEFUN(*parameter-name*)*
>   and set as a subgoal to code the relation calculated by this
>   function
>   and then write).

and

> IF   the goal is to code an argument
>   and that argument corresponds to a parameter of the
>   function
> THEN   write the parameter name.

The dotted lines in Figure 6.4 indicate the steps in the novice's problem solving which were combined in order to compile these production rules. An important point about knowledge-compilation is that a skill (or at least a component of a skill) can be acquired in a single learning episode. This was demonstrated by Anderson *et al.*, when they set the novice a further task to write a LISP program that returned the second element of a list, immediately after she had finished the first problem. They found that she was able to write a working function to return the second item in a list with only one attempt, whereas her first program had required a number of modifications to deal with errors. Anderson *et al.* argue that this occured because in writing her first function the novice compiled the rules shown above, which could then be applied directly to the task of writing a function to return the second item in a list.

The procedural stage

In the procedural stage, skills are practiced until they can be applied effortlessly. The production rules that are acquired in the previous phase of knowledge-compilation are strengthened by a process of *tuning*, where the speed and accuracy with which a rule can be applied increases with successful task performance. This is how automatic processes, to use Shiffrin and Schneider's (1977) terminology, are developed. Note the distinction in Anderson's theory between learning and practicing a skill: although one might acquire the rule necessary to undertake a task in a single episode, it may take large amounts of practice before the task can be completed with any recognizable degree of skill. This is particularly the case for some interfaces skills, such as touch-typing, where the potential for mis-application of similar but incorrect rules is very high. Thus, the one-trial learning exhibited by novice programmers contrasts with the repeated practice that is necessary to successfully perform some other skills such as touch-typing.

Automation of skills creates a problem in evaluating Anderson's account of skill acquisition, since a by-product of automation is that people are unable to verbalize the procedure with which they carry out a task. An expert is unlikely to be able to describe the automated rules that they use in performing their skill. This is indeed what is often found: experts are notoriously poor at expressing the methods by which they carry out tasks. Therefore, the production rules that underlie a skill cannot be observed directly but can only be inferred by researchers.

Transfer of procedural skills

So far we have discussed how examples and practice can lead to the acquisition of skills. Transfer from known tasks to similar tasks in a new context is another important mechanism for developing interface skills. ACT* offers an account for some of the phenomena associated with the transfer of skills. For example, production rules should be equally applicable to any two tasks that are structurally similar, regardless of any differences in the way they are described. Even if different notations are used to carry out structurally similar tasks, there should be *positive transfer* from one task to the other (i.e., the skills acquired on one task will facilitate the learning of another task).

Consider, for example, the task of learning to use a text editor. A number of text editors are available in the UNIX operating system. Some are line-based editors, in which the user issues keyboard commands for identifying the location of a relevant text block and then issues further commands which describe how the text should be changed. The text may then have to be re-formatted before the results of the editing operation are visible. Other text editors are screen-based, in that the user moves a cursor to the position on the screen where an operation is to be carried out and makes changes to the text which are visible as the user types them. Singley and Anderson (1985) investigated the extent to which users can transfer skills from line-based to screen-based editors. They compared the performance of three groups of users who learned to use the text editors over a six day training period: the first group used EMACS (a screen-based editor) throughout; a second group used either ED or EDT (both line-based editors) for four days before switching to EMACS for the last two days; and a third group used ED for days one and two, switched to EDT for days three and four, and then switched to EMACS for days five and six. The results of the experiment are shown in Figure 6.5. The faster edit times for the EMACS-only group throughout the training period demonstrate the advantage of screen-based over line-based editors. This difference probably reflects the immediate feedback provided

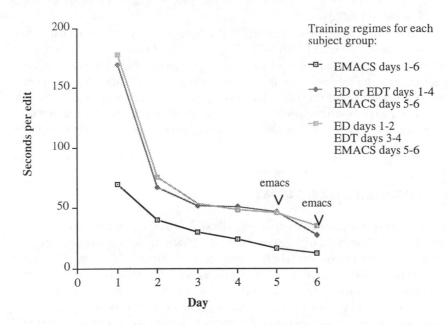

Figure 6.5 Results from Singley and Anderson (1985). Three groups of subjects received training under different regimes with line-based (ED and EDT) or screen-based (EMACS) UNIX editors. The absence of a difference on days 3 and 4 between the two and three-editor groups suggests positive transfer between the ED and EDT editors. However, on switching to the EMACS editor on day 5, performance in these groups was as poor as the EMACS-only group had been on day 2, suggesting no positive transfer between line-based and screen-based editors.

by screen-based editors, an advantage which we discussed in Chapter 5, and is not of immediate interest here.

More significantly, Singley and Anderson found almost total positive transfer between the ED and EDT editors when subjects switched from one to the other on day three of the training period. These text editors have completely different sets of command names, but nonetheless they are both line editors. Singley and Anderson argue that there can be positive transfer between line-based editors because the production rules acquired in practice with the ED editor are exactly the same as those that are required for using the EDT editor, despite the fact that carrying out edits in the two systems requires different keystroke-combinations. Practice with one editor acts as a form of practice for using another editor, because they may have different commands but they share essentially the same goal structure. As long as the same procedures are involved in using an interface there will be positive transfer between interfaces with different command sets. However, when the groups trained with line-based editors switched to EMACS on day five, their edit times were as slow as those of the EMACS-only group on day two. In other words, experience of line-based editors did not transfer positively to using the screen-based editor.

These results may seem surprising in the light of the negative transfer that occurs between alternative layouts of numeric keyboard, a point discussed earlier in this chapter. However, although subjects in Singley and Anderson's experiment appeared to compile rules with one editor, they did not receive extensive practice and therefore had not automated their skill. It may be that positive transfer between similar tasks which differ only in their command sets may not occur if the first task is practiced so extensively that it becomes automated.

Task-specificity and expertise

It is commonly assumed that expertise is flexible, that is, that experts can adapt their skills to changed circumstances. ACT* challenges this assumption, and seems instead to suggest that skills arising through knowledge-compilation are highly task-specific. For example, McKendree and Anderson (1987) gave subjects extensive practice in evaluating LISP expressions (that is, working out what value would be returned if a LISP expression were run on a computer). Subjects were first introduced to a number of commands, including CAR and CDR which we described above. These can be combined together in a shorthand notation to return items from different places within a list, such as

$$CADDR(a\ b\ c\ d) = c.$$

This expression finds 'the head of the tail of the tail of the list', which in full notation would be

CAR(CDR(CDR(a b c d))).

McKendree and Anderson found that the skills acquired in evaluating expressions like this did not transfer to the task of writing the same expressions. When subjects were presented with the task

"Given a list (a b c) write an expression which, when
evaluated, would return the answer c"

they were unable to supply the answer "CADDR(a b c d)", even though they had been trained to evaluate expressions such as this. McKendree and Anderson argue that evaluation and writing are essentially different procedures, despite the similarities between the tasks (both involve manipulating combinations of CAR and CDR functions). In other words, transfer of skills does not occur if different production rules are required, even if the tasks involve exactly the same declarative knowledge.

The results of this experiment may explain why instruction alone is often insufficient to teach complex skills such as computer programming. Textbooks focus on the declarative features of a task (e.g., what the purposes of CAR and CDR functions are, and how they should be combined), but they rarely describe *how* the task of programming should be carried out. When exercises are given in textbooks, they should require the reader to undertake a number of different tasks with the same information (e.g., writing, evaluating and debugging LISP expressions) so that procedures are developed for all the tasks that are encountered in programming.

An evaluation of ACT* as an account of skill

One of the attractions of ACT* is that it offers a comprehensive account of human skills. The ambitious nature of the framework nonetheless makes it a fairly easy target for criticism, of which four main areas can be identified:

ACT* as a realistic account of novice learning

In order that feedback is available for knowledge compilation to occur, novices must be consciously aware of the declarative knowledge they are using. However, there is evidence that people can acquire skills without ever being able to verbalize the declarative information involved in performing a task, or for that matter being aware that they have learned anything. For example, Berry and Broadbent (1984) found that subjects carrying out a process control task (which involved adjusting a sugar production system according to the size of work force available) showed clear improvements in performance which could only be explained by subjects having learnt a rule which related the variables together. However, subjects were unable to verbalize such a rule and were unaware that they were using one. Anderson's mechanism of knowledge compilation cannot account easily for learning without awareness such as this.

ACT* and errors

Another over-simplification of ACT* is that it suggests that novices do not learn directly from understanding their errors. Errors are seen in the ACT* framework to occur through the application of *mal-rules*, that is, faulty or inappropriate production rules. The application of a mal-rule will not lead to the successful completion of a task and so no learning should occur. Yet novices do appear to learn from their mistakes. More specifically, they use the feedback that errors provide in making decisions about what is the best solution to attempt next.

ACT* as a theory of expertise

The results of McKendree and Anderson (1987) suggest that expertise is task-specific. However, mere possession of task-specific production rules offers a somewhat limited explanation of the range of skills that an expert can develop, such as the sophisticated diagnosis and repair skills needed in the domain of process control and computer programming (see Payne, 1988, for an extensive discussion of this point).

The evidence for ACT*

Anderson makes extensive use of verbal protocols in his research into novice learning. However, in some instances his analyses do not stand up to close scrutiny. For example, Anderson *et al.* (1984) report the spontaneous decision of their subject (described above) to search for analogous examples to her current problem. Yet the verbal protocol given in the appendix to the paper clearly shows that she was prompted by the experimenter to search for examples. Our own experience of collecting protocol data (e.g., Ormerod and Jones, 1991) indicates how sensitive novices are to interventions by the experimenter, and also has shown how novice learning is considerably more complex than the simple declarative and knowledge compilation stages suggested by Anderson (though clearly the use of analogy in mapping examples from textbooks onto the current task is an important part of novice behaviour).

Despite these limitations, ACT* highlights a number of important phenomena to consider in developing optimum interfaces as well as methods for training interface skills. The next section examines these and other phenomena in relation to the training of procedural skills.

6.4 Training procedural skills

Choosing an appropriate training regime for a complex skill requires an understanding of both the characteristics of the task (e.g., whether it is repetitive or continuously varying) and the characteristics of the user (e.g., whether they are occasional or frequent users). For example, we discussed in Chapter 1 how the demands of safe operation necessitate extensive training for process operators and aeroplane pilots, so that the first action that comes to mind in an emergency is the optimum response. To achieve this, training should lead to the automation of skills for dealing with an emergency. The subject of training is too large to be covered in detail in this book, and for a recent and comprehensive review of training research the reader is referred to Patrick (1992). Nevertheless, some broad recommendations can be made for developing effective training methods.

The importance of feedback

Feedback is a prerequisite for the learning procedural skills, as it is for the performance of all interface skills (a point made in Chapter 5). In terms of the ACT* framework, feedback is necessary because knowledge-compilation only occurs when production rules are applied successfully in carrying out a task. If the execution of a production rule is unsuccessful, the learner must be informed of this failure in order to try and find another rule to apply. This is particularly important when learning a complex task, such as computer programming, in which a sequence of inter-related actions must be undertaken. Immediate feedback is essential on the results of each action as it occurs, otherwise it will not be clear which production rule had which effect.

For example, Ormerod and Jones (1991) describe the behaviour of a novice Prolog programmer trying to write a program to find the length of a list. The novice tested fourteen attempts before finally producing a working program, of which five were repetitions of previous attempts and six were modifications of an arithmetic expression which had originally been correct! Much of the blame for this painful learning episode can be attributed to the inadequate error messages given by the Prolog system. The most frequent message indicated an arithmetic error, which the novice understood to indicate a syntax mistake in the arithmetic expression. In fact the failure of the arithmetic occured only as a consequence of the real source of error, which was an inappropriate ordering of program lines. To overcome problems such as this requires the design of error handling systems which give useful and immediate feedback. Similarly, training methods must provide immediate and specific feedback about the cause and effects of each outcome and activity undertaken to complete a complex task.

Part-task training

One approach to training complex skills, such as driving, text-editing and programming, is to break them down into a number of components which are trained individually and then combined to give full task performance. This method of training, often referred to as *part-task training* (e.g., Stammers, 1982), can be interpreted in terms of the learning mechanism underlying ACT*. First, a complex skill may generate too large a memory load for a novice. Learning a task in smaller parts will lower this load, thereby reducing the number of errors that occur through forgetting. Second, if fewer production rules are necessary to perform a given task part, it will be easier to identify the effects of applying each rule, thereby making feedback more specific. Third, in learning later more sophisticated task components, subjects can utilize specialized rules compiled during the learning of earlier task components.

An experimental study which illustrates the benefits of part-task training was conducted by Mané, Adams and Donchin (1989). They trained subjects in a video game similar to Space Invaders, either by practice at the full task or by a part-task method which involved training first in button pressing, followed by spaceship recognition, followed by spaceship controlling. They found that subjects who received the part-task training prior to performing the whole task achieved a skilled performance criterion considerably faster than subjects who received full-task training. Part-task training led to superior performance relative to full-task training, not only for task components which received individual practice, but also for task components encountered only when the full sequence of tasks was finally undertaken. The superior performance of untrained task components resulting from part-task training might be an effect of reduced memory load, or may reflect the application of sophisticated production rules from the practiced components during the compilation of production rules for the new task components.

Minimal manuals

As suggested in Section 6.1, users often ignore instruction manuals, preferring instead to plunge straight into a task without the systematic use of instructional materials. For example, Carroll and Mack (1985) describe how people get themselves into situations in using word processors for the first time from which they cannot recover, either because they do not use the manual at all, or because they are unable to get the information they need when they do refer to the manual. If novices prefer to learn by immediately interacting with an interface, then this obviously makes writing manuals problematic. For example, it makes it less likely that novices will work systematically through a large set of exercises which are scheduled to introduce skills in a graduated sequence. Additionally, an extensive manual will generate a large working memory load, which is a source of novice errors (Anderson, 1987).

As a method for minimizing these problems, Carroll (1990) suggests that instruction should be focussed through the provision of a 'minimal manual', in which the materials given to novices are the smallest set that is practicable for them to learn how to perform the task. The method differs from part-task training in that a minimal manual does not decompose the user's task into components to be trained individually. Nevertheless, it is similar in that both approaches require that the cognitive load be minimized whilst a new skill is acquired. The concept of a minimal manual is based on three principles:

i.   Reducing the amount of material novices are faced with, so that they can start doing something useful without having to wade through reams of written material. This addresses the problems of large reading and memory loads faced by novices in learning a new skill;

ii. Providing a task orientation, in which instruction is focused on getting learners to perform tasks that are relevant to their needs as soon as possible. This addresses the problem that users want to perform something useful and obtain positive feedback as soon as possible in learning a new skill. It also minimizes the problems faced by novices in accessing specific information from manuals;

iii. Emphasizing methods of error recognition and recovery. This addresses the observation that novices turn to manuals only when they have made an error and are unable to see a solution to it.

Carroll, Smith-Kerker, Ford and Mazur-Rimetz (1987) describe an experimental study of learning a word-processing system which confirms the advantages of minimal over conventional self-instruction manuals. However, there may be drawbacks to the minimal-manual approach if it is used in the wrong learning context. For example, it does not support the re-use of prior knowledge (i.e., positive transfer) in learning new tasks. The emphasis is on action rather than understanding in minimal instruction. In effect, the message of a minimal manual is "Never mind how it works, just do it". On the other hand, one of the more subtle features of limiting the amount of information given to novices is that it forces the burden of explanation onto the novices themselves. For example, Black, Carroll & McGuigan (1987) found that novices trained with minimal manuals made more inferences, and thus demonstrated greater understanding of a word-processing system, than novices trained with a complete and exhaustive manual. The role that reasoning plays in developing interface skills is examined in the next chapter.

The concept of minimal instruction might also be applied to the design of interfaces, particularly in deciding how big a command set to give novices. For example, the Wordstar word processing package provides a display which can be altered according to level of expertise. Novice users are expected to see a full screen display of command names, intermediates a reduced set, and experts hardly any commands. The idea is that, as one learns the command names, the need for a screen-based reminder is reduced so that more of the screen can be taken up by the document and less by the commands. The opposite approach is adopted in Microsoft's word processing package, which displays short menus as the default mode, but can be extended to show full menus when a more complex command set is required. In this system the idea is to restrict the complexity that novices must deal with, which is more in accordance with a minimal approach to instruction. In truth, this is a somewhat spurious analysis, since the command dialogue of Wordstar gives rise to a memory load which is absent, or at least reduced, in the menu-based dialogue of the Word graphical user interface. Nevertheless, the choice between minimal or exhaustive presentation of information provides a good example of the context-sensitivity of interface design that we alluded to in Chapter 1. Minimalism is fine when the user needs to make only limited demands on the range of tasks that an interface supports. However, an exhaustive presentation of information may be necessary where a large range of novel interface tasks must be supported.

6.5   Conclusions

In this chapter we have concentrated on the interface skills which users develop for carrying out procedures. Some skills are so extensively practiced that they become automated and highly task-specific, whilst others require conscious attention whilst they are performed. The positive side of this is that automated processes have smaller demands to make on the cognitive system, so that the user is able to carry out other tasks at the same time. The negative side is that automated skills are inflexible and prone to error if interrupted. Procedural skills are acquired in a process of learning-by-doing through rote mapping of examples and solutions onto the current task.

There are many implications that follow from learning-by-doing for the transfer and training of procedural skills. A skill that has been acquired, but which has not been extensively practiced, will transfer to a new situation as long as the procedures for conducting the tasks are sufficiently similar. On the other hand, there will be no transfer between two tasks which require different procedures, even if they involve the manipulation of exactly the same interface features (e.g., evaluating and writing LISP programs). Three approaches to training procedural skills are the provision of immediate feedback, part-task training, and the use of minimal instruction. Each of these approaches is intended to minimize cognitive load whilst the skill is being learnt.

Whilst ACT* is useful in providing an framework around which to organize our understanding of procedural skills, it is vulnerable to a number of criticisms. For example, the description of learning as occuring through analogical mapping without conceptual understanding is an oversimplification and ignores individual differences in strategy. It is clear that restricting an account of interface skills to rote-rehearsed procedures offers a somewhat impoverished account of both expert and novice behaviour. The next chapter examines the roles that conceptual understanding and prior knowledge play in determining the successful use of an interface.

CHAPTER 7

Skill as understanding

Overview

In this chapter we focus upon the analytical components of interface skills. The chapter begins by examining the use of prior knowledge in understanding interfaces. We describe the concept of 'schemas', a type of psychological theory that describes the way in which experts store and apply their prior knowledge in a form that is abstract and generalizable to novel tasks and contexts. A 'plans' account of programming expertise is discussed as an example of a schema theory. We then introduce 'mental models' as an alternative approach. A mental model is a simplified mental representation of a task, which helps a user to predict appropriate responses. Arguably, the accuracy of the user's mental model of an interface determines their success in performing novel or complex tasks with it.

We also address the role that prior knowledge can play in learning and training new interface skills. The use of conceptual metaphors in the design of interfaces, in particular the 'desktop' metaphor, is examined. We then discuss the potential advantages and disadvantages of guiding learning through the elicitation of metaphors based on existing knowledge. An alternative approach is to train novices with a novel conceptual model to help them organize their understanding of an interface. The utility of this approach to training interface skills is evaluated, and is compared with the provision of training programmes based on examples or formal instruction.

7.1   Expertise as analytical skills

In the last chapter we discussed how skills which are based on automatic procedures can be applied without conscious thought and do not place heavy task demands on the user. However, there are many advantages to possessing skills for *analysing* the interface, that is, skills which are based on understanding rather than rote performance. Analytical skills are like the controlled processes discussed in the previous chapter, in that they are highly flexible but require conscious thought before application. They allow the user to understand how a task is performed with one interface, which may enable them to generalize their understanding to another interface and to modify aspects of their performance when the desired results are not obtained. This chapter examines the nature, acquisition and training of analytical skills.

Analytical skills place an emphasis on the flexible use of declarative knowledge in reasoning rather than the rote use of automated procedures (see Chapter 6 for a discussion of declarative and procedural knowledge). There are two main aspects to analytical skill: first, analytical skill requires prior knowledge; and second, the prior knowledge used by experts is abstracted to some degree, or at least generalizable across a number of specific instances.

Abstract and concrete knowledge of interfaces

In order to understand the role that prior knowledge plays in interface expertise, it is necessary to investigate the ways in which experts' knowledge differs from that of novices. Prior to studies of expert users of computer interfaces, the first experts to be studied in detail were chess grand-masters (e.g., De Groot, 1965). De Groot challenged the commonly-held belief that grand-masters have an ability to plan a long way ahead and calculate the consequences of a number of possible moves before choosing the best option. Instead, he argued that the skill of grand-masters is based on studying thousands of games over a number of years, and then being able to recall an appropriate move for a particular board position from memory. In other words, the skill of grand-masters appears to rely more on memory than on analysis.

Chase and Simon (1973) demonstrated the importance of memory for chess-game positions in an experiment in which they presented novice and expert chess-players with boards showing either realistic or randomized chess-game positions. They measured the number of glances at the board that subjects required to reconstruct the positions accurately, and found that, whereas grand-masters were able to reconstruct positions from realistic games within one or two glances, novices required as many as twenty glances. However, novices and grand-masters required an equally large number of glances to reconstruct random positions. This suggests that grand-masters were able to recall the whole board layout from long-term memory as a single chunk of information, whereas novices had to remember all of the positions of individual chess pieces. The ability to recall chess-game positions during a game of chess is a great advantage, since recall of a chunk of knowledge about a particular layout provides the grand-master with information about the best move to make for the current situation.

Similar studies have been carried out to examine the nature of experts' knowledge about computer programming. For example, McKeithen, Reitman, Reuter and Hirtle (1981) investigated the recall by novices and experts of

ALGOL programs, the lines of which had either been scrambled or were left in a realistic order. Figure 7.1 illustrates the experimental materials they used. Like Chase and Simon's study of chess players, they found that experts recalled more lines from the normal programs than novices, but that there was no difference between experts and novices in the recall of scrambled programs. However, whereas Chase and Simon found that the advantage for experts over novices declined with repeated presentations of the same chess-game layout, McKeithen *et al.* found that the difference between experts and novices actually increased with repeated presentations of the same program. This may occur because, unlike chess, where identical layouts are encountered repeatedly in the experience of a grand-master, an expert programmer will rarely encounter two programs which are perceptually identical. Therefore, whilst it makes sense for grand-masters to remember the exact chess-game positions they have encountered previously, it would not be sensible for programmers to remember the exact lines of code that make up a program.

The most important point about the study of McKeithen *et al.* is that it shows how programming expertise is not based solely on *concrete* aspects of

```
Section of the normal version:

      BEGIN  PTR: = I;
         WHILE  ORDERS (TRIALS + 1 , PTR) ≥ J  DO  PTR := PTR + 1;
            ASSERT  PTR < 27;
            FOR I:= 1  UNTIL  TRIALS + I  DO
            BEGIN  TEMP(I) : =  ORDERS (1,J);
               ORDERS (1,J): =  ORDERS (I, PTR);
               ORDERS (I, PTR) := TEMP (I)
            END
      END

Same section as it appeared in the scrambled version:

         WHILE  ORDERS (TRIALS + 1 , PTR) ≥ J  DO  PTR := PTR + 1;
      BEGIN  PTR: = I;
            FOR I:= 1  UNTIL  TRIALS + I  DO
            END
         ASSERT  PTR < 27;
      END
            BEGIN  TEMP(I) : =  ORDERS (1,J);
               ORDERS (I, PTR) := TEMP (I)
               ORDERS (1,J): =  ORDERS (I, PTR);
```

Figure 7.1 Section from the normal and scrambled programs used by McKeithen *et al.* (1981) in their study of novice and expert ALGOL programmers.

programming code (i.e., the exact lines of code). Instead, expert program-mers appear to store relatively *abstract* knowledge about programs, that is, knowledge about general classes and components of typical programs, whilst they forget the exact words and layout of specific programs. In a task which involves recalling a program exactly, an expert would not have access to an exact representation of a similar program stored in long-term memory, but would have to reconstruct the presented program over a number of trials using their abstract knowledge to guide them.

Further evidence of the abstract nature of some interface expertise comes from studies of command knowledge. For example, UNIX has an enormous set of available commands, perhaps too large to be remembered item-by-item. Indeed, it has been argued that hardly anyone becomes totally proficient in UNIX to the extent of knowing the full command set, and even experts make errors and have gaps in their knowledge (Draper, 1985). Nonetheless, experts are able to make up for incomplete knowledge by generalizing what they do know about UNIX commands. Doane, Pellegrino and Klatzky (1990) have found that experts organize their knowledge of UNIX at an abstract level con-cerned with the shell system, whereas intermediates and novices organize their knowledge at a concrete level concerned with the names of individual commands. An abstract organization of UNIX commands is advantageous because it allows experts to generalize their knowledge to novel tasks. Even when experts do not know the specific syntax that is needed to undertake a new task, they can use their abstract knowledge to predict the kind of syntax that would carry out the task. Thus, an important aspect of interface expertise is the ability to make comparisons between familiar and unfamiliar tasks according to some shared characteristic at an abstract level.

Differences in the ways in which experts and novices organize their knowl-edge have also been observed in programming tasks. In a second experi-ment, McKeithen *et al.* (1981) trained novice and expert programmers in a keyword sorting task, and found that, whilst experts grouped keywords that were conceptually related (e.g., *for, next* and *go*, which are all concerned with flow of control), novices grouped keywords according to irrelevant syntactic similarities (even resorting to groupings as desperate as *bits, of,* and *string*). Similarly, Adelson (1981) has shown how experts and novices recall pro-grams in different ways, experts tending to recall in bursts of semantically-related information and novices tending to recall in bursts of syntactically-related information. Thus, experts appear to develop an abstract mental representation of programming information.

Schema theories of expertise

If experts make use of abstract knowledge in performing novel and complex tasks, then a theory of expertise should describe how abstract knowledge is

represented in memory and then retrieved in a useful form. The most ubiq-
uitous psychological explanation of memory for abstract knowledge is the
schema (after Bartlett, 1932). Schemas are mental representations in which
past experience is organized hierarchically, so that information about higher-
level (or more abstract) concepts is inherited by lower-level concepts. Each
concept in a schema has *slots* for matching with new instances which are
encountered in the task domain. They also contain *default values* that consist
of information which is relevant to most situations in which the schema is
evoked, as well as *unique values* that describe information which is specific to
particular situations.

A schema theory of programming expertise

An example of a schema theory is the *plans* theory of Soloway and colleagues
(e.g., Soloway, Erhlich and Bonar, 1982; Soloway and Erhlich, 1984),
designed to account for programming expertise. This theory is a generaliza-
tion of Van Dijk and Kintsch's (1983) theory of text comprehension. Van
Dijk and Kintsch propose that the content of stories can be understood at a
number of levels of abstraction (from overall plot down to the structure of
individual sentences). In a similar way, Soloway proposes that programming
expertise consists of a set of hierarchically organized plans at various levels
of abstraction which can be used to comprehend and write computer
programs.

Figure 7.2 shows an example of a Pascal program in which a number of
plans can be identified (Erhlich and Soloway, 1984). The program reads in a
series of numbers until a pre-specified value is read, and then calculates the
average of the series. Erhlich and Soloway identify three main types of plan
within this program: a 'running total loop plan' which runs a loop to input
and process each number, 'variable plans' which initialize and update variables (the program contains variable plans to implement both a 'counter'
and a 'total sum'), and a 'skip guard plan' which checks for specific values
that must be excluded.

In the plans theory, programming expertise does not consist simply of frequently encountered chunks of code, in the way that chess expertise appears
to consist of knowledge of exact board layouts that occur during games.
Instead, it appears to consist of abstract structures which chunk together
related information about programming concepts. Interestingly, the observation of McKeithen *et al.* (1981), that the difference between experts and
novices in the recall of realistic programs actually increased by over a number
of trials, may be explained by plans theory. Experts would, over a number of
trials, come to recognize plan-like structures in the realistic program, which
they could then use to reconstruct the program.

Figure 7.3 illustrates two of the plans described above in terms of their

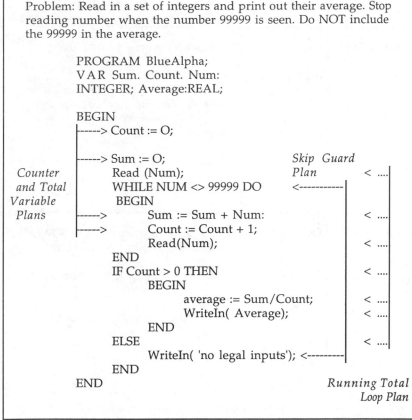

Figure 7.2 A sample program from Erhlich and Soloway (1984) showing three types of programming plan which represent expert knowledge about the program.

schematic components (concepts, slots, default and specific values). Programming plans are usually represented using a specific kind of schema representation known as a *frame* (Minsky, 1977). The process of programming consists of evoking a relevant plan and then adapting its contents to the specific programming problem (by filling slots, using default values, etc.). As the program in Figure 7.2 suggests, plans are not serially ordered in a program, but are interwoven. Thus, another part of the programming task is to fit the components generated by each plan together. Rist (1986, 1991) has extended the plans theory to explain how plans might be interwoven. He proposes that each plan has a 'focal line', that is, a key line of code which suggests the main properties of the rest of the plan. For example, the focal line in the variable plan is probably the line X = X + Y (where X and Y are slots to be filled by specific values), because the arithmetic suggests the purpose of the

Plan	Concepts	Slots and fillers (plain = defaults, italics = special values)
Counter variable plan	Description	Counts occurences of an action
	Initialization	Counter := 0 *Counter := 1 for programs which use count as denominator in division*
	How used?	Counter := Counter + 1
	Type	Integer
	Context	Iteration
Running-total loop plan	Description	Builds up a running total in a loop *Counts the number of iterations*
	Variables	Counter, Running total, New value
	Set up	Initialize variables *Pass existing count from another program component*
	Action in body	Read, Count, Total

Figure 7.3 A description of two programming plans in terms of their schematic components (after Detienne, 1990).

rest of the plan. In Rist's theory, experts construct their code by first producing all the focal lines of relevant plans, and then fleshing out the remaining parts of the plans around the focal lines.

Programming languages may be easier to learn if the syntax clearly delineates plan-like structures. For example, the notations for looping constructs in Pascal (the *For..next*, *While..do* and *Repeat..until* notations) serve as 'beacons' in programs which can be used to comprehend the code (Rist, 1986; Wiedenbeck and Scholtz, 1989). Not all languages provide beacons to guide programmers to the structural properties of programs. For example, Green (1989) suggests that Prolog provides few notational features which enable the programmer to recognize plan-like structures within the code. This may be one of the reasons why Prolog can be a very difficult language for novices to learn (Taylor and du Boulay, 1986).

Programming plans also offer a potential approach to developing support tools such as program editors. For example Johnson and Soloway (1985) have constructed a computer-based system called PROUST which teaches Pascal programming by presenting novices with exercises designed to emphasise expert plans. Similarly, the Programmer's Apprentice (Waters, 1985) is a tool with plan-like templates which experts can then adapt to fit their current programming task. Whether either of these systems is successful at training novices or supporting experts has yet to be evaluated. Whilst tools might support the use of programming plans, they can also violate them. For example, Green (1989) describes how a speech input system for constructing Pascal programs was a poor medium because it required a strictly linear input of program code, a restriction which failed to support the plan-order in which experts might construct their code.

Recently, the status of plans as a general explanation of programming expertise has been challenged. For example, Gilmore and Green (1988) found that, while cueing plan-like structures by indentation is an effective aid to experts' comprehension of Pascal programs, similar cues do not assist in the comprehension of Basic programs. They suggest that plan-like knowledge may be specific to Pascal. An alternative explanation for the absence of plan-like knowledge of Basic has been suggested by Davies (1990). He found that Basic programmers who had received formal training in design, in which they were encouraged to adopt a principle of top-down and breadth-first program development, showed more plan-like behaviour than programmers of similar proficiency who had not been formally trained in design.

Ormerod and Ball (1993) have suggested that plan-like behaviour emerges whenever program design is carried out in an hierarchically-structured fashion, and that the appearance of plans may be simply be a by-product of design strategy rather than reflecting knowledge of abstract program structures. They also provide further evidence against a significant role for plans in Prolog programming. They show how two experts managed to produce exactly the program, yet the the order in which each programmer generated their code was completely different. This observation seems inconsistent with the notion of focal-line expansion of plans developed by Rist (1986, 1989).

The general utility of schema theories

We have illustrated schema theories with reference to the domain of programming, but they are widely regarded as a useful approach to describing some components of interface expertise and they have been applied to interface tasks other than programming. For example, in studies of novices and experts carrying out text-editing tasks, Kay and Black (1985) found a trend towards increasingly sophisticated types of text-editing behaviour with greater expertise, which is consistent with the development of plan-like knowledge structures. Studies like this offer the potential to inform interface

designers about appropriate command sets for text-editors to support different levels of expertise.

Some general concerns should be raised about schema theories as an
account of interface skills. First, schemas describe an efficient way of storing
relevant knowledge but they do not necessarily explain how experts apply
that knowledge. Second, there is a tendency in the literature on human-
computer interaction to use the term 'schema' when all the writer really
means is 'knowledge'. Unless the specific properties of schematic structures
are outlined, describing expertise as based on a schema representation
carries no additional meaning. Finally, our critique of the programming
plans theory may also generalize to other schema theories (see also Alba and
Hasher, 1983, for a detailed critique of schema theories).

Expertise as mental models

Another approach to describing the ways in which prior knowledge is used to
understand interfaces is that of *mental models*. In its most general form a mental model, like a schema, is a mental representation of a problem and its
solution. However, mental models differ from schemas in that they are seen
as analogues of the real-world. In other words, a mental model has the same
functional nature and structure as the system that it simulates.

Consider, for example, the workings of an automated teller for dispensing
bank notes. Payne (1991) examined the answers given by subjects to questions about the way in which an automated bank teller functions, in order to
examine the mental models that people construct of the workings of such
machines. He found a wide variation in the beliefs that people hold about
how these machines function, which he attributes to the formation of different mental models. These models many be incomplete or incorrect, but they
are usually sufficient to enable users to complete the tasks for which the
machine is designed. Two examples of the mental models reported by Payne
are shown in Figure 7.4. The first model indicates that subject 15 believed
that the machine itself carried out all of the processing functions and the
bank card stored the personal data. The second model indicates that subject
14 envisaged a central computer which stored the personal data and carried
out the main processing functions, the local machine acting to handle the
inputs and outputs of the dialogue only (a model which is closer to a correct
description of most systems).

Payne suggests that the quality of users' models can determine their performance with the machine. For example, he cites the following part of an
interview:

"E: What happens if you type ahead during pauses?"
"S: I wouldn't have thought it would take it in – it's concentrating on
    something else."

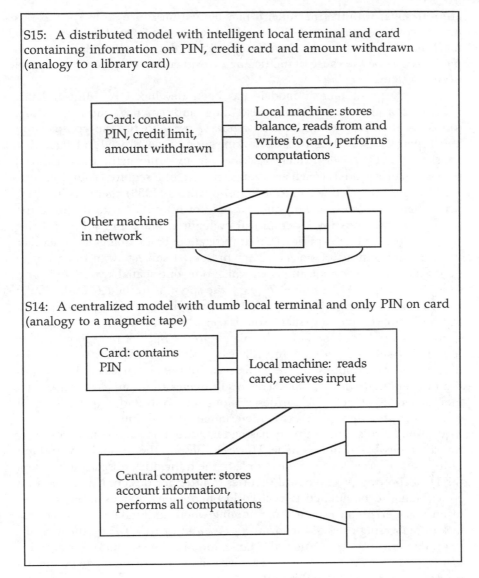

S15: A distributed model with intelligent local terminal and card containing information on PIN, credit card and amount withdrawn (analogy to a library card)

Card: contains PIN, credit limit, amount withdrawn

Local machine: stores balance, reads from and writes to card, performs computations

Other machines in network

S14: A centralized model with dumb local terminal and only PIN on card (analogy to a magnetic tape)

Card: contains PIN

Local machine: reads card, receives input

Central computer: stores account information, performs all computations

Figure 7.4 Mental models of an automated bank teller, inferred from subjects' answers to questions about the way the machines operate (Payne, 1991).

This seems to suggest that the subject has a mental model in which the exchange of machine prompts and user inputs is seen as a conversation, each waiting for the other to finish before an appropriate response can be made. In fact, most systems allow the user to enter a string of inputs without pauses between them, the prompts being provided merely as guidance for users who are unfamiliar with the machine. It may be that misconceptions such as this occur because of an erroneous mental model people form about bank teller machines. If this is the case, then it suggests that the mental models that people hold may be the cause of inefficiencies (and customer queues) in the use of such systems.

The concept of mental models has been developed by Johnson-Laird (1983) as a theory of human reasoning and language understanding. Johnson-Laird's theory outlines a *procedural semantics* for the construction and testing of mental models in working memory. A procedural semantics consists of a set of rules which describe how task information is organized into alternative models which are then searched for a required solution.

Although a detailed account of Johnson-Laird's (1983) theory is inappropriate here, it is worth giving one example of his research to illustrate the theory and methods for its empirical evaluation. Mani and Johnson-Laird (1982) conducted an experiment to investigate peoples' ability to remember the relative positions of objects. They presented subjects with determinate descriptions of objects' positions, in which only one spatial layout could reasonably correspond to a sentence (e.g., 'the spoon is to the left of the knife; the plate is to the right of the knife'), and with indeterminate descriptions, in which more than one spatial layout is consistent with a sentence (e.g., 'the spoon is to the left of the knife; the plate is to the right of the spoon'). They found that subjects recalled the relative positions of objects best for the determinate descriptions, but verbatim recall of the sentences was better with the indeterminate descriptions. This result suggests that subjects construct a mental model of determinate descriptions, but resort to rote learning of sentences where they give rise to indeterminate descriptions.

Johnson-Laird's (1983) theory has had considerable success in accounting for laboratory-based phenomena. However, despite the undoubted strength of his work, it provides a good example of the difficulties (discussed in Chapter 1) of applying psychological theories to interface design. Nevertheless, it is beginning to be applied to real-world phenomena. For example, novices often make errors in Prolog programming when undertaking the necessary task of rephrasing rules of the forms *if p then q* and *p only if q* into the logically equivalent form *q if p* (Ormerod, Manktelow and Jones, 1993). Ormerod *et al.* argue that errors occur because subjects fail to create an adequate mental model of the rule to be rephrased.

Mental models theory also has some interesting implications for learning and training interface skills. For example, the research of Mani and Johnson-Laird (1982) suggests that when subjects form a mental model, they encode only a simplified set of information. This suggests that, in situations where a verbatim recall of information is required (such as operator responses in emergency procedures), training based on conceptual overviews of the system should specifically be avoided since these will encourage subjects to form a mental model of the system. Instead, methods which encourage rote recall should be encouraged.

Mental models in research on human-computer interaction

The use of the term 'mental model' has been the source of much confusion in human-computer interaction. This has occured because there are a number of possible perspectives from which models of an interface might be described. These perspectives are illustrated in Figure 1.3 (Chapter 1), which shows a representation of possible views of the user interface, adapted from Young (1983). The term 'mental model' is best used in human-computer interaction to describe the model or models that a user has of an interface, though it has also been used (wrongly) to describe the designers' model of the user's knowledge. The classification of users' models and their various synonyms can be confusing, because different views of interfaces are studied for different reasons. Lansdale (1985) offers three categories which describe how the term mental models is used by interface designers:

Conceptual representations

These are mental models which offer an overall view of a system, and are synonymous with analogies (e.g., the typewriter/word processor analogy). The calculators analysed by Young (1983) and discussed in Chapter 4 illustrate the role of conceptual representations. For example, the ALG calculator is the simplest calculator for novices to learn, because it requires a conceptual understanding of keying-in orders which is exactly consistent with the order in which arithmetic expressions are normally written. The RPN calculator is the most difficult to learn because it requires the user to learn a novel conceptual representation (a 'stack') to describe the order in which the keying-in must be undertaken. However, having formed a conceptual representation based around a model of a stack, the user then has a highly flexible understanding of the device which aids in the undertaking of complex calculations. Young describes conceptual representations such as the stack model of the RPN calculator as 'surrogate models', since they serve as simplifications of the conceptual structure of the devices that they describe.

Implied dialogue models

These refer to the models which users infer about the functionality of an interface. Lansdale (1985) suggests that meaningful command names (as discussed in Chapter 4) encourage users to infer models about the structure of function of an interface (e.g., 'archive' implies a particular filing system or structure). The perceptual appearance of an interface can also give rise to an implied-dialogue model (e.g., the visual effect of shrinking into a folder that accompanies the closing of windows displaying the contents of a disk directory). In this sense, an implied dialogue model represents the *affordance* of an interface, a concept discussed in Chapters 4 and 5.

Cognitive models

These are mental models that a user gains through experience of a system (i.e., the knowledge that makes a user reflect either novice or expert under- standing) and are closest to the sense in which Johnson-Laird (1983) uses the term 'mental model'. Payne's (1991) description of how even experi- enced users may possess an inadequate or faulty mental model of automated bank tellers is an example of this type.

One reason why progress has been slow in applying the concept of mental models to interface design is because a user's mental model must be inferred by the researcher in the light of performance on an interface task. The num- ber of possible mental models that a person might hold about a device like a calculator or an automated bank teller is infinite. Therefore, to infer a specif- ic model from observations of users' performance is necessarily a subjective exercise. This contrasts with the laboratory-based research of Johnson-Laird (1983), in which a finite set of mental models that might rationally be created can be predicted independently of task performance. A distinction can therefore be made between the formal psychological use of the term, in which the mental models that people reason with are predicted *a priori*, and the looser use of the term found in applied literature, in which a user's mental model of a device is inferred from their task performance.

For example, Payne (1991) infers the mental models that subjects might hold from studying their answers to questions about the interface. He is care- ful to try and distinguish between answers which may reflect that a new infer- ence is being made in the light of the question, and answers which reflect a genuine pre-existing mental model. However, there is considerable room for interpretation in placing subjects' answers in either category.

Interestingly, Payne observed that subjects sometimes gave anthropomor- phic (machine = sentient being) descriptions of machine function, such as the example given above – "I wouldn't have thought [the machine] would take it in – it's concentrating on something else." Subjects used anthropomorphisms such as this when they wanted to structure their descriptions of the machine's behaviour, but did not do so when supporting direct inferences or explana- tions. In other words, anthropomorphism is useful for describing abstract phenomena but not for drawing specific inferences. This indicates a general problem faced by researchers using interview data to examine subjects' mental models: the generation of 'models' may be a communicative rather than an inferential behaviour. In other words, people may use models to describe a complex idea to another person, but this does not necessarily mean that they use the model to make decisions about what action they should carry out.

A second problem with the concept of mental models is the failure of some researchers to distinguish between the mental models that people have in their heads and the provision of external models to help people reason. The assumption is that, if you provide people with an external model to help them learn a task, they will internalize this as their mental model. A related error is to assume that if subjects' performance improves as a result of presenting them with an external model, then this model must reflect the knowledge that they already have in their heads (e.g., Kieras and Bovair, 1984). If the presentation of an external model aids subjects' performance, then it is possible that they did *not* already possess a similar mental model, since had they possessed such a model their task performance would not be enhanced further by the provision of the same external model.

Mental models have been the focus of much research interest (e.g., Gentner and Gentner 1983; Rogers, Rutherford and Bibby, 1992). However, we believe that there is a limited amount that can be said about mental models that is of direct benefit to interface designers. The rationale for psychologists studying mental models is obvious: they are interested in theories of how minds do and do not work. However, it is not obvious at this stage that the study of mental models can tell interface designers much beyond the observation that users make predictions about how an interface works. The important thing for designers is knowing how to put this insight into practice, an issue which has been a focus of Chapters 2–5.

7.2   Learning through understanding

Understanding the causes of failure

While it may not be possible to determine the exact representation of expert knowledge (e.g., whether it reflects the use of schemas or mental models), it is clear that some form of conceptual knowledge of interfaces is acquired and used. We have already outlined in Chapter 6 how Anderson's (1983) ACT* theory describes a method of learning which relies on the compilation of procedures. Knowledge compilation is a method in which learning occurs only as a result of successful task performance. However, it is unlikely in anything but the most simple and repetitive tasks that one simply keeps trying different things without questioning one's performance or learning anything about the task until an attempt succeeds. If we try something and it fails, then we want to know why it has failed. The acquisition of conceptual knowledge seems to require a method of learning which is more analytical and evaluative.

An example of such a method is embodied in Reisbeck and Schank's (1989) theory of *case-based reasoning*. Case-based reasoning is a form of schema theory in which prior knowledge is used to generate and test predictions about novel situations that are related to familiar instances (or *cases*). The theory offers learning-through-failure as an alternative to the success-based mechanism of ACT*. Reisbeck and Schank propose that human memory (and therefore learning) is dynamic, in that the performance of new tasks is undertaken through the generation and testing of predictions based on prior knowledge. If these predictions are confirmed, then the task is performed successfully and nothing new is learned. If the predictions are not confirmed and task performance fails, then the reasons for the failure are traced until the causes are identified, and new predictions are added to the knowledge held in memory. Case-based reasoning has perhaps its greatest influence in the development of computer-aided learning systems (e.g., Du and McCalla, 1991). The theory is as yet under-specified and is arguably even more difficult to test empirically than ACT*. Nevertheless, the idea that people think about the task they are learning and try to evaluate the reasons for their success or failure seems intuitively reasonable.

If people generate and test predictions in order to learn new interface skills, then it is important to understand the sources of information that predictions can be based upon. Information sources include the interface itself, the provision of conceptual models, the use of analogous examples to solve new tasks, and the presentation of formal instruction. These information sources can be provided as part of training regimes, and are discussed in Section 7.3. An alternative approach to explicit training, the elicitation of prior knowledge which is of relevance to a novel task or interface, is discussed next.

Metaphors for interfaces

Part of the process of understanding a new task is to identify a conceptual framework around which to organize knowledge of the various components of an interface. In Chapter 4 we discussed how consistency in the design of command languages makes the concepts that underlie interfaces easier to learn. Another way in which consistency can be achieved is to provide an organizing principle onto which users can map different components of the interface. Organizing principles often take the form of a *metaphor*, that is, a description of a simple non-computer task which can serve as a model for a more complex or novel computer-based task.

Perhaps the best known conceptual metaphor for an interface is the 'desktop', developed for the Xerox STAR (Smith, Irby, Kimball, Verplank and Harslem, 1982) and subsequently widely used in graphical user interfaces. The desktop metaphor simplifies the explanation of file handling and disk operation by presenting computing concepts to users in familiar language which is relatively free of computer jargon. The metaphor operates by eliciting prior knowledge of familiar concepts to represent computer objects, such as the placing of paper documents (files) in folders (sub-directories) which are placed in windows (directories), and which can be dragged around to different windows on the desktop (moving or copying files from one directory to another) or thrown away by placing in the trash can (deleting a file). The use of superficial familiar features, such as an icon shaped like a trash can which bulges when something is thrown away, reinforces the metaphor as well as providing feedback to the user.

The desktop metaphor is actually more complex than simply a desktop = file-directory metaphor, since it relies on WYSIWYG and WIMP interface features (described in Chapter 3). Some of these features add additional metaphors to a simple desktop metaphor (e.g., who ever heard of having windows on a desktop?). Other features actively violate aspects of the desktop metaphor (e.g., dragging a file from one window to another window represents the copying rather than movement of a file to a new disk). Despite these subtle complexities, the success of the desktop metaphor can be judged by its widespread adoption for modern personal computer interfaces.

When metaphors fail

Metaphors that elicit prior knowledge of everyday tasks rely for their success on achieving a close match between objects in the everyday task and objects in the interface. When the match is not exact, there is a risk that people will use their knowledge inappropriately in trying to understand the new interface. In some cases, an inexact match between metaphor and interface is not problematic. For example, we have already seen how certain aspects of the desktop metaphor do not match the interface itself (e.g., copying rather than moving files as a result of dragging them across windows), yet these do not seem to be a source of difficulty for users.

Other mismatches do appear to cause problems, as in the use of a 'typewriter' metaphor for word processor interfaces. Intuitively, many aspects of the task of word processing are similar to the task of typing, notably the use of a keyboard to produce text. However, word processors treat text in a fundamentally different way to typewriters, in that text is not a physical entity which is fixed in position by each keystroke. Douglas and Moran (1983) identified a number of errors in learning to use a word processor which were attributable to the use of a typewriter metaphor. For example, they observed novices holding keys down for too long, an action which leads to unwanted repetition of keystrokes. Also, novices were often inefficient in their deletion of text, using the backspace key rather than by using block-deletion commands. Similarly, the properties of the Return key were often misunderstood when it was equated with the carriage return of a typewriter (the Return key functions as a text-block separator rather than simply an end-of-line).

Despite these observations, the extent to which even poor metaphors (like typewriter = word processor) are a significant problem is open to question. For example, Allwood and Eliasson (1987) found that, whilst inefficiencies were attributable to the use of a typewriter metaphor, errors tended to be caused by subjects making inappropriate analogies between command functions within the word processing system itself. The subjects in Douglas and Moran's study were trained secretaries who presumably had extensive typing experience. Errors made by experienced typists might therefore reflect the mis-application of automated procedures (e.g., hitting the Return key at the end of every line). The subjects in Allwood and Eliason's study were students who had familiarity but not expertise with typewriters. They did not have automated procedures (such as pressing the Return key at the end of each line), and were therefore less likely to make typing-related errors than the experienced secretaries. However, the students' knowledge of typing might encourage them to use inefficient methods (e.g., using the backspace key to delete blocks of text) to save the effort of learning new word processing commands, since they might have perceived such learning as having value only for the duration of the experiment.

Results like this raise doubts as to whether people really use metaphors to help them learn new tasks, and suggest that people do not rely entirely on metaphors as the basis of their knowledge. Conceptual metaphors are directed at novices, and are intended to allow them to gain an initial understanding of a system without requiring instruction or experience in all its details. However, possession of a partially effective metaphor may discourage the novice from learning new features of a system which require them to conceptualize the task in a novel way (e.g., treating text as an object upon which commands can be implemented rather than as a physical entity whose properties are fixed when it is produced). Whilst a metaphor may give the novice sufficient confidence to undertake the task of learning a new interface, it does not remove the need for instruction or experience.

7.3   Training users' analytical skills

Training with conceptual models

The conceptual metaphors discussed in the previous section evoke prior knowledge that might usefully be applied to understanding an interface. However, for many interface tasks there is no real-world familiar equivalent. In this case, novices can be be trained in a novel conceptual model of the task. The provision of a conceptual model has been described in psychological literature as equivalent to giving an 'advance organizer' (Ausubel, 1968), in that a simple overview is provided for a new task. Advance organizers can have a large impact on the effectiveness of training. For example, Figure 7.5

The procedure is quite simple. First you arrange items into different groups. Of course one pile may be sufficient depending on how much there is to do. If you have to go somewhere else due to lack of facilities that is the next step; otherwise, you are pretty well set. It is important not to overdo things. That is, it is better to do too few things at once than too many. In the short run this may not seem important but complications can easily arise. A mistake can be expensive as well. At first, the whole procedure will seem complicated. Soon, however, it will become just another facet of life. It is difficult to foresee any end to the necessity for this task in the immediate future, but then, one never can tell. After the procedure is completed one arranges the materials into their appropriate places. Eventually, they will be used once more and the whole cycle will then have to be repeated. However, that is part of life.

Figure 7.5 The 'launderette' passage used in Bransford and Johnson's (1972) experiment which demonstrated enhancement of recall by an advance organizer.

shows a passage used in a memory experiment conducted by Bransford and Johnson (1972). They found that recall of the passage was much better if subjects were told beforehand that the topic was washing clothes. However, providing the topic after the passage had been read did not improve recall compared with that of subjects given no summary of the topic. This experiment demonstrates that the ability to learn new information is dependent on being able to place it in the context of what we already know. If no context exists (e.g., where the topic was not known before presentation of the passage in Bransford and Johnson's experiment) then learning is very poor. Thus the intention of providing a conceptual model of an interface is to give an advance organizer around which a novice can easily assimilate new information.

Mayer (1975, 1976) investigated the efficacy of providing a conceptual understanding of the computer for training novices in BASIC (the conceptual model is shown in Figure 7.6). Although knowledge of the internal workings of the computer is not essential to the task of programming in BASIC, Mayer found that provision of the model led to faster and more error-free learning, particularly for tasks which were novel or of greater complexity. In other words, the conceptual model appeared to give subjects an ability to generalize their knowledge to new tasks and make predictions about ways to perform, an approach suggested by Reisbeck and Schank's theory of case-based reasoning discussed above.

An experiment by Caplan and Schooler (1990) suggests that the advantages of conceptual models may be more limited that the results of Mayer's experiments imply. They examined the performance of novices learning to use a computer drawing package. They found that training with a conceptual

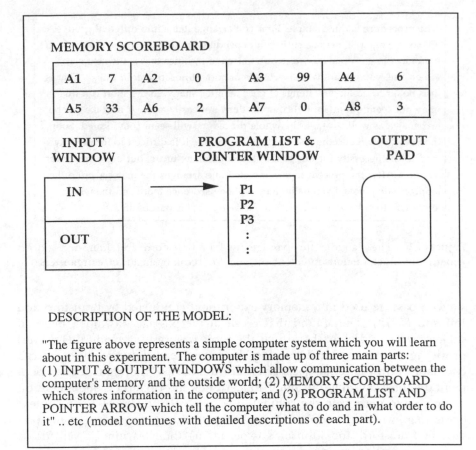

Figure 7.6   Part of Mayer's (1975) concrete model of a Basic-like language.

model facilitated performance in complex drawing tasks (i.e., those which involved a number of tools selections and manipulations to complete a picture). However, training with a conceptual model impaired performance in simple tasks (i.e., those which involved only one or two steps to complete the picture) relative to that of subjects who were not presented with a model. Thus, conceptual models are useful for complex tasks which involve a degree of problem-solving before they can be undertaken. However, if subjects are encouraged to develop a conceptual of a device, it can distract them from compiling the kinds of automatic procedures that are sufficient for executing simple tasks.

There is clearly an incompatibility between training by conceptual models and by minimal instruction (Carroll, 1990), an approach which was introduced in Chapter 6. While one approach extends and elaborates the information provided to the user, the other approach strips away as much of the detail as possible. In any instance, the approach to adopt will be determined by the nature of the task to be trained. Where a task involves the learning of rote action sequences (e.g., simple text editing) a conceptual model is unlikely to prove beneficial. On the other hand, where a task involves a large degree of intellectual activity (e.g., programming) a conceptual model appears more useful.

Examples and instruction in training

We have seen in Chapter 6 how knowledge-compilation, a process by which task-specific procedures are acquired, uses a method of analogy to map the solutions of example problems onto new problems. In Anderson's (1983) theory of skill acquisition, analogy is a procedure for mapping of examples onto new problems which does not necessitate the development of conceptual understanding. A number of researchers have suggested that analogical reasoning also provides a mechanism for the acquisition of conceptual knowledge (e.g., Gick and Holyoak, 1983; Holland, Holyoak, Nisbett and Thagard, 1986). However, there appear to be significant limitations on peoples' spontaneous ability to carry out successful analogical reasoning. For example, Reed, Dempster and Ettinger (1985) found that analogical transfer was effective only from a more difficult analogous problem plus solution to a simpler task. This finding has important implications for the development of training regimes.

Training to transfer strategic skills

The limitations of analogical reasoning and their impact on training are illustrated in an experiment carried out by Ormerod, Manktelow, Robson and Steward (1986) to investigate novices' reasoning about simple Prolog programs. Many Prolog textbooks use a program for identifying family relationships to illustrate important concepts such as the structure of program statements, the search for alternative solutions to queries, and the definition of facts and rules. The example also has the apparent benefit of being easy to understand. However, family relationships have two features, a highly familiar content (everyone knows about brothers, uncles, cousins and so on) and a diagrammatic representation in the form of a tree structure, which make them a somewhat suspect example to use. Ormerod *et al.* investigated the efficacy of this example by manipulating the thematic content of the task (familiar or unfamiliar) and representation of the relationships between individuals (list or diagram), and measuring subjects' ability to find solutions to program queries with each set of materials. The materials and questions used in their experiment are illustrated in Figure 7.7.

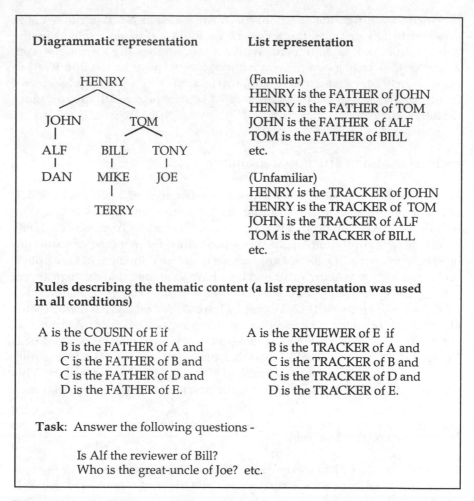

Diagrammatic representation

List representation

(Familiar)
HENRY is the FATHER of JOHN
HENRY is the FATHER of TOM
JOHN is the FATHER of ALF
TOM is the FATHER of BILL
etc.

(Unfamiliar)
HENRY is the TRACKER of JOHN
HENRY is the TRACKER of TOM
JOHN is the TRACKER of ALF
TOM is the TRACKER of BILL
etc.

Rules describing the thematic content (a list representation was used in all conditions)

A is the COUSIN of E if
    B is the FATHER of A and
    C is the FATHER of B and
    C is the FATHER of D and
    D is the FATHER of E.

A is the REVIEWER of E if
    B is the TRACKER of A and
    C is the TRACKER of B and
    C is the TRACKER of D and
    D is the TRACKER of E.

Task: Answer the following questions -

    Is Alf the reviewer of Bill?
    Who is the great-uncle of Joe? etc.

Figure 7.7 The task and materials used in the experiments by Ormerod *et al.* (1986, 1990).

They found that subjects were fast and accurate in reasoning with familiar rules and a diagrammatic representation of individual relationships. Subjects were also accurate, albeit very slow, when reasoning with unfamiliar rules (a hypothetical management system) where the individual relationships were shown as list of text rather than as a diagram. When subjects were given either familiar rules and listed relationships or unfamiliar rules with diagrammatic relationships, they made many more errors although response times remained fast. There are two surprising aspects to these results: first, that performance with familiar content should be so inaccurate when relationships were represented as a list rather than as a diagram; and second, that response times should be so much faster with unfamiliar content when relationships were represented as a diagram rather than as a list.

The results of this experiment suggest that two strategies were used for carrying out the task: a perceptually-based strategy which uses prior experience of the diagrammatic displays of family trees to identify family relationships by their layout; and a rule-based strategy in which individual relationships are mapped step-by-step onto rules which describe the content. The presence of either familiar rules or a diagrammatic representation encouraged subjects to adopt a perceptually-based strategy. However, this strategy was only successful with materials having both familiar rules and a diagrammatic representation. It failed (though it led to fast response times) when one of these features was absent.

In a similar experiment, Ormerod, Manktelow, Steward and Robson (1990) found that subjects who were trained initially with the unfamiliar-list materials made significantly fewer errors when they later received the familiar-list and unfamiliar-diagram materials. It appears that subjects transferred the rule-based strategy which had been used successfully with the unfamiliar rules plus text-based list of relationships, thereby overriding their natural inclination to be swayed by the presence of familiar content or diagrammatic representation. Prior training with familiar-diagram materials did not have the same positive effect on subsequent performance with familiar-list and unfamiliar-diagram materials, since it only encouraged further the adoption of an inappropriate perceptually-based strategy.

These experiments indicate that examples and exercises for training materials should be chosen with care. Novices are likely to be influenced by superficial aspects of a task, and are unlikely to grasp the underlying conceptual structure of the task unless they are forced to adopt a slow and painful strategy for dealing with the task materials. This seems to suggest that textbooks and manuals should only use examples which are unfamiliar to the novice. This idea would appear counter-intuitive to most educationalists, since the point of using familiar examples (like conceptual metaphors and advance organizers) is to enable the novice to relate new concepts to their existing knowledge. In practice, training examples should be chosen which are comprehensible but which are not overly familiar. Of course, this requires the instructor to make a subjective judgement about the familiarity of the training materials.

Training and formal instruction

If analogical transfer sometimes encourages the development of inappropriate strategies, then what about formal instruction in the underlying principles of an interface task? For example, a formal explanation of the difference between paragraphs, document sections and pages might enhance the

learning and application of style-setting commands of modern word processing software. However, of the three word processing systems we use in our
laboratories, none of them presents any formal instruction on this topic.
Instead they rely on step-by-step examples from which the user is required to
infer general principles of document segmentation and the application of
styles.

The effectiveness of formal instruction has not been the focus of much
empirical work in human-computer interaction, perhaps because it seems
intuitively obvious to designers that novices will find it complex, confusing and
unhelpful. Evidence from studies of logical reasoning (e.g., Cheng, Holyoak,
Nisbett and Oliver, 1986) suggests that training in principles of formal logic
does not lead to sustained improvements in reasoning performance. On the
other hand, Fong, Krantz and Nisbett (1986) found that formal instruction in
statistical reasoning did improve performance. Thus the efficacy of formal
instruction in training appears to be dependent on the task domain.

One interface task in which formal instruction has received study is computer programming. Conway and Kahney (1987) investigated the training of
recursive programming skills in SOLO (recursion being the conceptually difficult concept of defining a programming function in terms of a smaller subset of itself). Initially, they examined subjects' ability to use examples to solve
new programming problems without formal training, and found that novices
made superficial mappings between examples and programming problems,
and failed to make the necessary deep analogical mappings to understand
their recursive structure. These results are consistent with the finding of
Ormerod *et al.* (1986) who showed that the selection of a reasoning strategy
can be influenced by superficial task features.

Conway and Kahney then carried out an experiment to investigate which
kind of training regime would facilitate the acquisition of recursive programming skills. The four regimes they investigated were:

i.   Examples-only training;
ii. Examples plus a formal description of recursion;
iii. Examples plus explicit mappings which showed what the common
     elements were in examples and target problems;
iv. Examples plus explicit mappings plus a formal description of recursion.

They found that only the last regime was effective in training novices to produce correct solutions to target problems. In other words, novices were
unable to use either explicitly mapped-out examples or formal instruction on
their own, but required both to form an effective conceptual understanding
of recursion. Thus it appears that there are no easy solutions available to the
instructor in training highly complex skills such as recursive programming.
Instead, a rich set of information sources, including conceptual models,
examples and formal instruction, is required to build up a sufficiently
detailed conceptual representation of a complex task.

7.4   Conclusions

This chapter has examined the analytical nature of interface skills. Such skills may be based on the application of abstract knowledge, as is suggested by schema theories of expertise, or on the mental simulation of an interface's function, as is suggested by a mental models account. Conceptual understanding of new interfaces may be encouraged through the provision of conceptual models or through solving problems by analogical reasoning.

There are methodological and theoretical limitations to skill-as-understanding as a complete account of interface skills. Primarily, it is impossible to *know* what representation of an interface a user holds in their heads. Researchers can only *infer* mental representations through empirical studies. As a result, the representations offered are often simply descriptions, or at least abstractions, of the interfaces themselves. Another problem facing accounts of skill-as-understanding is that much of what users do seems to follow from them observing the interface itself, rather than from the application of conceptual knowledge about the interface. Theories of interface skill, therefore, need to determine how much of the user's behaviour can be explained in terms of their conceptual knowledge, and how much follows naturally from the structure of the interface itself. The use of external information in the interface to inform and guide users' skills is discussed in the next chapter, where we examine the extent to which skill is based on exploration.

CHAPTER 8

Skill as exploration

Overview

In this chapter we examine a view of user skill which focuses upon the ways in which users extract information from an interface. Under this view, much of what comprises expertise is *strategic* knowledge, in that sophisticated strategies are required to explore interfaces effectively. We discuss *display-based problem solving*, a theory which captures this notion. The focus in the second section of this chapter is on how interfaces can be designed which allow the user to explore them and acquire strategic knowledge without first having to learn how to interact with the interface. In particular, the concept of direct manipulation is examined, and its potential advantages and disadvantages are discussed. Finally, we discuss the implications of skill-as-exploration for training interface skills. The emphasis is on user-led training, and on the development of interfaces to support this. In particular we look at the concepts of hypertext, multimedia and virtual reality, and we look at how to train users to search effectively.

8.1   Introduction

The development of interface technologies such as graphical user interfaces has allowed interface designers to offer more than just the efficient support of existing tasks (such as word processing, accountancy and programming). Instead, they can enable users to carry out new or existing tasks in novel ways. Indeed, the CHI '90 conference on Human Factors in Computer Systems had as its overall theme the idea of 'empowering people'. Behind this cliché lies the notion that interface design should not simply be about de-skilling the tasks that are carried out using computers. Instead, designers should use technological advances to enable people to carry out tasks that were not previously considered. There are three main aspects to empowerment: first, people should be given powerful interfaces with which to undertake new tasks; second, people should be given interfaces which match the tasks they want to carry out with minimal learning; and third, interfaces should be designed to train people in novel and interesting ways.

Consider, for example, the task of computer programming. In Chapter 4 we described research by Green (1977) into the design of programming notations. Whilst this research might make the conventional task of programming easier, it does not change the task itself. Programming remains essentially a task of instructing the machine in a sequence of steps that must be followed in order to undertake the required procedure. Many computer scientists (e.g., Hoare, 1981) want to change the focus of programming from a task of machine instruction to a task of designing novel solutions to problems. This change of emphasis from coding to design is as an important impetus to the development of object-oriented systems, a technological development which is discussed in Section 8.2.

Some programming problems may be best served by developing specific languages to tackle them. For example, Wilde and Lewis (1990) have developed NoPumpG (illustrated in Figure 8.1), a system which links graphics system to a spreadsheet. They argue that a graphical spreadsheet is a natural programming medium for creating simulations, since it allows a greater

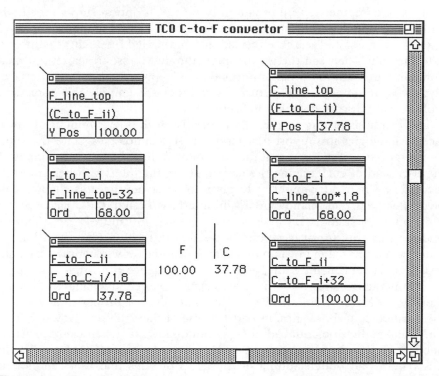

Figure 8.1   The NoPumpG programmable spreadsheet of Wilde and Lewis (1990).

degree of experimentation than conventional programming languages. NoPumpG exemplifies a trend towards domain-specificity in interface design. All-purpose programming, graphics, word processing and numerical packages are gradually being superseded by application-specific systems which support particular tasks for specific user-populations.

An important consideration for the design of interfaces which extend users' capabilities is the need to minimize learning and training efforts. In other words, users should be able to undertake tasks without having to learn a new set of commands or actions to control the interface. Instead they should carry out the task using essentially the same kind of actions that they would use if they were performing the task directly and not through the intermediary of a computer interface. The concept of *direct manipulation* embodies this idea, in that users carry out actions using the mouse and pointer system of a WIMP interface which mimic the actions that are required to carry out the task as a physical action. For example, dragging files across from one window to another mimics the activity of physically moving objects around. Direct manipulation, a concept discussed further in Section 8.3, empowers users by allowing them to interact with interfaces in a way that is natural to them.

Another example of empowerment is to be found in the design of novel sources of computerized information. Advances in optical storage methods and video-digitizing technologies have enabled the development of creative ways to represent and search information. This has led researchers to investigate the match between the way people naturally choose to search information, and the way in which information is represented at an interface. Hypertext, multimedia and virtual reality systems exemplify this approach, and these are discussed further in Section 8.4.

The emphasis in this chapter, therefore, is on the development of interface technologies to support new tasks. These technologies share the common properties of trying to put the control of an interface into the hands of the user, and of encouraging the user to *explore* the interface when interacting with it. The reader might be forgiven for worrying that this chapter simply reflects current 'bandwagons' in interface design that may well be supplanted by newer and better technologies in future years. This is undoubtedly true (or at least, one hopes that this will be the case), but we reiterate a point made in Chapter 1 that the fundamental cognitive capabilities of users remain the same, and as a consequence the issues that affect users' interaction with these interfaces remain largely unchanged. To date there is a dearth of empirical evidence that allows a proper evaluation of these interface technologies, in part because of their novelty. However, one can evaluate their potential advantages and disadvantages, an exercise which needs to be undertaken with any emerging interface technology. To do this requires an understanding of the kinds of skills that users employ in exploring interfaces.

8.2   Expertise as exploring interfaces

Display-based problem solving

Users of interfaces rely to a large extent on learning skills in the forms of rote-learned procedures (as described in Chapter 6) and conceptual models (as described in Chapter 7). There is, however, another source of information for users: *the interface itself.* In other words, users can search the interface for information about how to use it and what to use it for. Larkin (1989) describes this kind of expertise in a theory of *display-based problem solving.* Like Anderson's (1983) ACT* (described in Chapter 6), skill is based on the use of production rules to undertake the actions that are necessary to interact with an interface. Also like ACT*, Larkin's theory proposes that short-term working memory represents task information against which production rules are matched. However, unlike ACT* which considers only the internal representation of information, in Larkin's theory there are two repositories of information: an internal register of the user's goals; and an external representation of information in the task environment.

Larkin (1989) uses the Towers of Hanoi problem as an example of the way in which the external display is searched for information that contributes to successful problem solving. The problem is illustrated in Figure 8.2. The external display is a three-dimensional model of the disks and pegs. In solving this problem, a person can use information about the relative sizes of disks and their positions to determine which disk to move next. They do not need to plan a set of moves in their head in order to solve the problem. Instead they can read from the display which disk should be the next to be moved. Thus the problem is solved, not by the user planning a sequence of moves and then executing them, but by the user selecting a disk which appears to be in the wrong place and moving it closer to the correct place. If a disk cannot be moved because it is covered, then the user must move another disk first. The important point is that it is the external display, not the individual solving the problem, which determines the goal-structure of problem-solving, that is, the order in which different actions should be undertaken.

There is some evidence that manipulating the display can facilitate performance with this problem. For example, Zhang (1991) found that solution times were faster and errors were less frequent when the disks and pegs problem was replaced with a version using cups of water. In this version, the restriction on not placing smaller cups over larger cups was explicit in the display (smaller cups would fall into larger cups!). Thus external constraints made it easier for subjects to select which move to make next, because the rules governing legitimate moves were explicitly represented in the problem.

One of the implications of Larkin's theory is that, with the acquisition of expertise, people acquire strategic knowledge which facilitates the use of displays as external repositories for information. For example, Davies (1990)

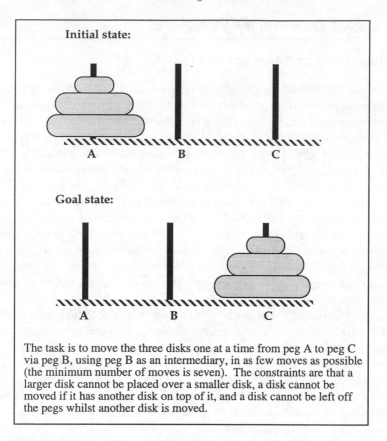

The task is to move the three disks one at a time from peg A to peg C via peg B, using peg B as an intermediary, in as few moves as possible (the minimum number of moves is seven). The constraints are that a larger disk cannot be placed over a smaller disk, a disk cannot be moved if it has another disk on top of it, and a disk cannot be left off the pegs whilst another disk is moved.

Figure 8.2   The Towers of Hanoi problem.

found that expert programmers tended to rely on externalizing their code earlier and did so more frequently than novices, who tended to rely more than experts on developing their solutions mentally before transferring them to the VDU screen. Davies (In Press) examined the role that external displays play in programming skill in a series of experiments, in which features of the programming environment were systematically disrupted. For example, in his second experiment, expert and novice programmers were required to produce code using a modified editor which did not allow lines of code to be revised once they had been created. The effects of this restriction were far greater for experts than for novices, suggesting that experts place a greater reliance than novices on the task of revising their externally-represented code. Novices, on the other hand, relied more on developing their solutions using internal memory, and so had essentially finished developing a line of code once it was committed to the editor.

The external environment can be seen to play a significant role with a number of different interfaces. For example, in Chapter 4 we described an experiment by Mayes *et al.* (1988) which showed that expert users of a word processing system were no better than novices at recalling the screens and menus of the system. This is an example of apparent reliance by experts on the interface display. Using a menu-based word processing system does not consist simply of applying rote-learned procedures. Nor does it necessarily require an understanding of the principles of organization of word processing commands. Instead, users actively search the display for the commands they require. In Chapter 4 we saw that Howes and Payne (1990) described how users often adopt a strategy of 'menu-flicking', because a graphical interface makes it more cost-effective for users to flick quickly across all the menus in search of an appropriate command than it is to learn the command names and keystrokes.

The notion of display-based problem solving is being applied in an increasing number of areas. For example, Wu and Anderson (In Press) discuss the influence of displays on the transfer of programming skills, and Mannes and Kintsch (1991) have also applied the notion to text-editing tasks. There remains much work to be done to determine the nature and extent of expert strategic knowledge, in particular whether there are any general display-based strategies or whether strategies are specific to particular interfaces. Whatever the future is for theories of display-based expertise, they indicate that expertise does not necessarily entail learning every detail of an interface. Instead, designers must develop interfaces which support the acquisition and use of strategies for searching their displays. For example, Sharples and O'Malley (1988) have proposed the design of a computer-based writer's assistant, which serves both as an *aide memoire* to externalize decisions made during the planning of a text, and also as a way of restructuring the text plans, so that alternative views of the same material may be created.

The reductionist nature of research into human skill tends to encourage a false dichotomy between the different views of skill presented in this chapter and Chapters 6 and 7. Elements of all three types of skill, procedural, knowledge-based and exploratory, are observable in most interface tasks. For example, programmers clearly acquire some procedural skills, as can be witnessed in their ability to write down on paper familiar sequences of code in the absence of a computer display. There is also clearly a role for conceptual knowledge in enabling expert programmers to develop programs for new problems. Nevertheless, the notion of display-based problem solving seems to capture the active search of the display that experts seem to carry out. Similarly, there appear to be elements of both rote-learned procedures and conceptual understanding in the performance of expert word-processor users. However, word processing systems with graphical interfaces also encourage users to develop strategies for searching the display economically.

Object-oriented systems

If some of the skill that experts bring to using interfaces rests in their use of displays, then there is a need for systems which encourage experts to develop and utilize effective display-based skills. This may be one reason why object-oriented systems have been the focus of much recent attention, particularly for interfaces such as graphics drawing packages, CAD/CAM systems and programming languages. The basic idea behind object-oriented systems is that, rather than having to specify routines, procedures or functions, the user can describe the *objects* that exist in the problem domain and then specify their attributes. In particular, the user of an object-oriented system with a graphics-based interface can focus upon manipulating objects on a screen rather than specifying procedures for carrying out tasks. The basic characteristics of an object-oriented system are described in Table 8.1.

Consider, for example, a graphics-based drawing package. Figure 8.3 shows an example of a load-carrying crane drawn using Aldus Intellidraw. In this drawing, the lines that make up the crane components have been linked together such that each of the components represents a single graphic object. These objects can be assigned properties, such as symmetricality, foreground or background status, and so on, which enable their production, alteration and reproduction to be controlled to a far greater degree than would be possible in a simple line drawing system. As Figure 8.3 illustrates,

Table 8.1 The main characteristics of object-oriented systems.

i.	Using an object-oriented interface consists of defining and assigning attributes to objects which represent either physical or conceptual entities that exist within a problem domain.
ii.	Objects may have actions associated with them, such that when an object is activated the action is carried out. For example, clicking on a 'button' such as might be found in a Hypercard stack, as illustrated in Figure 8.4, will cause the action that is associated with that button to occur (in this case, movement to another stack).
iii.	Objects are activated by messages being passed to them, either from the activation of other objects, or from user or system inputs (e.g., pointer selection mechanisms).
iv.	Objects can be grouped together in classes which share common attributes, such that when the attributes of one object are changed, the attributes of all in the same class are changed. In some systems, objects are organized hierarchically, such that attributes or actions associated with a high-level object are inherited by objects at lower levels.

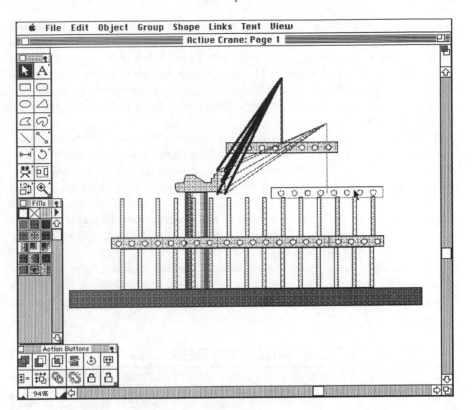

Figure 8.3 An example of a drawing created using Aldus IntelliDraw, an object-oriented graphics package. The drawing components are defined as objects which can be linked to each other. When one object is moved (e.g., the girder held by the crane), objects which are linked are adjusted to fit with the movement (hence the appearance of an animated crane arm when the girder is moved).

the user can simulate the movement of beams using the crane arm by clicking and dragging the crane arm. Objects, once created, can then be added to an object style sheet and reproduced in other drawings. If one of the attributes is then changed, it will change in all the occurrences of that object.

The concept of object-orientation is perhaps at its most powerful when applied to programming environments. One of the simplest of these is Hypercard, which is a graphics-based programming system (though it can also be used to create hierarchical indexes of text and graphics without recognisable programming effort). An example of a Hypercard stack is shown in Figure 8.4. Hypercard essentially reverses the traditional order of programming, in which the program functionality is developed first and a graphical front end may then be attached (Green, 1989). When programming in Hypercard, the first act is to create graphical objects which are then given the required functionality by the addition of 'scripts' to particular objects. Scripts are essentially small programs which are activated when an object receives a message, and they determine what action should then occur. Since the initial focus in Hypercard programming is on creating the display, the programmer then adding the required functionality, it offers a particularly attractive programming environment for interface designers (at least as a prototyping tool).

Whether or not a graphics-based system is used, there appear to be advantages for object-oriented programming languages over conventional languages. From a computer-science perspective, they offer a method of producing modular programs which can be maintained and tested in small independent chunks. They also allow for the incremental development of programs, thereby reducing the scale of the program design task at any one time. From a psychological perspective, they may also change the programmer's conception of the task of programming.

Object-oriented programming represents a switch in focus from a linguistic activity to a problem solving activity. For example, Rosson and Alpert (1990) suggest that an object-oriented approach to software design may be

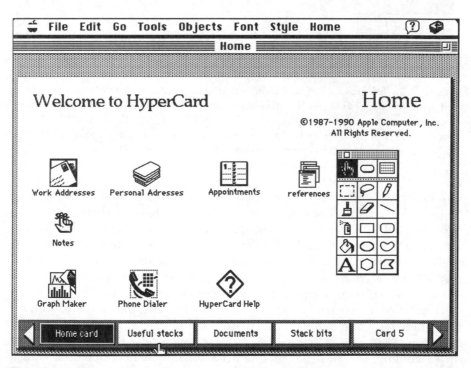

Figure 8.5 The 'home' card of Hypercard, illustrating the use of buttons for navigating to other stacks, and a tear-off pallette of drawing tools.

advantageous for some domains in which the problems are themselves represented in terms of objects (e.g., components of architectural designs). However, there are some domains in which the concept of objects with associated attributes and actions does not map onto the nature of the problem (e.g., routine accounting procedures for manipulating numerical data). Nevertheless, Rosson and Alpert argue that even when a problem domain cannot naturally be conceived in terms of objects, applying object-orientation as a metaphor can improve the quality of design solutions and encourage novel designs to be considered. They illustrate this with the following example:

> "... consider the design of a wastebasket in an object-oriented office system. Design might begin by characterizing the state/behavior of wastebaskets in the real world (e.g., their typical shape and size and the facts that they accept waste ...). But application of the general metaphor of intelligent communicating objects can quickly lead to an understanding of the problem that goes well beyond the attributes of wastebaskets in the real world. Problem analysis might proceed by analogy to the intelligent objects the designer knows best – human beings – by trying to imagine what behavior a person might exhibit if assuming the wastebasket role in an office. Thus, a wastebasket might be primed to look for certain kinds of deposits, and send out a warning to some other object, if one appears ..." (Rosson and Alpert, 1990, p. 365).

The point that Rosson and Alpert are illustrating is that thinking about design in terms of objects rather than procedures encourages the designer to think creatively. In this example, the designer is encouraged to role-play, by considering the kinds of activities they could usefully carry out if they were given the task of acting as a wastebasket. This might encourage the designer to consider the development of 'intelligent agents' in their interface, which monitor the user's performance and try to predict what kinds of action or feedback the user would find most useful. The use of intelligent agents in interface design is currently a subject of much research interest (e.g., Adelson, 1992).

Object-oriented systems for graphics, programming and other applications appear to have many attractions. The question is, do they live up to this promise? The commercial success of object-oriented graphics packages is testimony to their utility. Mohageg (1991) has shown that performance with object-oriented graphics systems are superior to bit-mapped systems (in which each individual pixel is treated as a discrete component of the drawing) in most drawing tasks (though bit-mapped systems are still necessary for painting tasks).

Evidence for the advantages of object-oriented programming is less clear. There is some evidence that an object-oriented approach to design may have advantages over traditional problem decomposition methods. For example, Kim and Lerch (1992) found that solution times were reduced and the amount of mental simulation of designs that was required was considerably reduced when designers used an object-oriented design methodology. However, their observations were based on a pilot study of only three subjects.

Although object-orientation may be a good approach to program design, are undoubtedly problems in learning languages for object-oriented programming. For example, Eisenberg, Resnick and Turback (1987) found that some concepts of object-oriented programming, particularly the requirement to treat procedures as objects, were sources of difficulty for novice programmers. Similarly, Nielsen, Frehr and Nymand (1991) found that, although novices were able to learn the basics of programming in Hypercard very quickly, they had difficulty in dealing with the concept of inheritance. Our own experience of using and teaching Hypercard suggests that the ability to create graphics of objects and then assign functionality to them may be a mixed blessing. Whilst it allows users to achieve tangible designs quickly, their Hypercard stacks are invariably poorly structured and therefore hard to comprehend and debug. This is a particular problem in Hypercard, since the program code is distributed across a number of objects, and so related code cannot easily be inspected. Thus, whilst object-oriented programming languages may encourage designers towards creative programming, they may also encourage a lack of rigour which leads to problems in maintenance.

8.3   Learning through exploration

One of the implications of research into display-based problem solving (e.g., Larkin, 1989; Draper and Oatley, 1990) is that the emphasis on learning should be minimized if interfaces are designed effectively. In other words, the user should not need to acquire knowledge or skills related to the interface itself: they can simply use it. There are two prerequisites for a system to be usable without initial learning: first, it must be self-explanatory; and second, there must be a method of interacting with the system that is consistent with existing skills. The first prerequisite has already been discussed in Chapters 4 and 5, where we examined Norman's (1988) notion of affordance. Here we examine *direct manipulation* as a method of interaction that requires little or no initial learning.

Direct manipulation interfaces

The basic idea behind direct manipulation is as follows: the user should be able to participate in an interface dialogue which mimics the actions that would be required to perform a task if it required the manipulation of physical entities. Thus it is *direct* to the extent that actions are carried out on interface objects themselves rather than through an intermediary command language. For example, Figure 8.5 shows a process control plant simulator designed using LabView. The user can simulate the control of this plant by moving levels up and down the tanks, turning on pumps and shutting valves, and so on. This kind of system could potentially be used as a control room interface or as a training simulator.

Table 8.2 gives a list of properties described by Schneiderman (1987) that a well-designed direct manipulation interface should have. In truth, they are properties of any well-designed interface. Direct manipulation interfaces are, where the input and output technology are sufficient, becoming a *de facto* standard for interface designers. There is a strong relationship between object-oriented systems and direct manipulation, since the power of the object-oriented metaphor is realized most strongly when objects can be manipulated as tangible artifacts.

There are a number of potential problems with direct manipulation (see, for example, the critique offered by Hutchins, Hollan and Norman, 1986). Table 8.3 outlines five areas where the concept might be challenged. Three

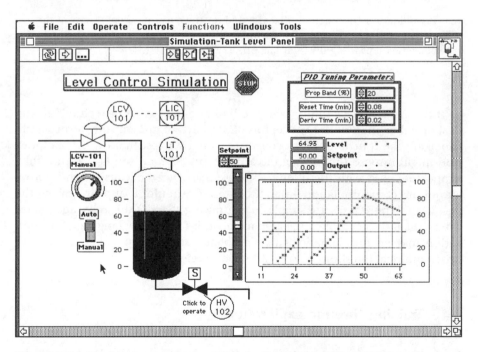

Figure 8.5 A process control operation simulated in LabView which has a direct manipulation interface. The controls (e.g., auto/manual switch) can be adjusted as analogues of physical controls using the mouse and pointer.

Table 8.2   Properties of direct manipulation (and other well-designed) interfaces as described by Schneiderman (1987)

i.	Explicit and visible action: the user points and manipulates objects, thus making it explicit what is being operated upon.
ii.	Immediate feedback and modeless interaction: When an object is manipulated the effects of the action should become immediately apparent. In text editors with separate 'insert' and 'edit' modes, errors often occur because of failures to change mode when the activity changes.
iii.	Incremental effects: Actions should parallel users' movements, so that the user can break a complex task into smaller components which have a summative effect. The effects of altering the plant shown in Figure 8.7 can be viewed without 'running' a simulation after each change.
iv.	Intuitive interaction - the interface must match the user's model of the task. Direct manipulation interfaces generally require the simulation of real world tasks. The aim of this is to enable operation of an interface without prior learning (i.e., don't think about it – do it !).
v.	Learning by onion peeling: This describes the idea of gradual revelation of system complexity. The user should explore the more complex parts of an interface only when their knowledge is sufficient for them to do so.
vi.	Pre-validation - only actions that are valid for the current task and state of the system should be possible (for example, as used in greying out of menu and dialogue items to signify their inappropriateness).

of these are relatively minor, and are limitations of current technology and design rather than of the concept itself. Perhaps of greater long-term significance are the fourth and fifth concerns, which suggest that direct manipulation might impose the burden of carrying out tasks on the user rather than supporting tasks with the interface. These issues arise because of the central aim of direct manipulation – that the interface should disappear, leaving the task itself. Where tasks are simple (e.g., drawing shapes) this is advantageous: where the task is complex, the absence of an intermediary language for interaction may be problematic. These problems are only speculative, but they raise empirical questions worthy of further exploration.

8.4   Training through exploration

If direct manipulation can be used effectively in interface design, and the problems raised in the previous section can be overcome, then is there really a role for training of interface skills? One should distinguish between the skills necessary to use an interface (e.g., using the functions of a computer-based document outliner), and the skills necessary for carrying out a task (e.g., planning the structure of a complex document). The aim of interface design should be to reduce the need to learn the former so that the user can concentrate on the latter. Clearly, there are some skills, such as programming, which will always require a significant amount of training. Simply providing a direct-manipulation interface does not remove the need for programming skills, since the programmer must still make decisions about the purpose and nature of the program. In other words, changing the interface may alleviate problems with coding, but it cannot address completely the difficulty of problem understanding and program specification. Nevertheless, an opportunity arises from the notion of skill as exploration to use interfaces creatively for training users in new domains.

Table 8.3   Potential problems with direct manipulation interfaces

i.   Repetitive tasks – Whilst incremental effects break complex tasks into manageable components, users may have to repeat the same actions again and again where the task is repetitive. A well-designed system needs a macro or scripting mechanism to handle repeated manipulations.

ii. There are problems in dealing with the concept of variables and with distinguishing individual objects from a class of objects (e.g., in visual programming). Direct manipulation requires that objects are tangible entities, yet this does not fit well with the concept of a variable as a slot that can represent other objects.

iii. The pointer system is not necessarily accurate enough or sufficiently controllable for some tasks which require accuracy. The advantage of issuing commands is that one can give explicit coordinates.

iv. Direct manipulation, by its very nature, requires a close mapping between existing knowledge and interface skill. This might restrict interface design to tasks that users already do, and prevent the design of new tasks. A corollary of this is that users may restrict their use of interfaces to highly familiar routines. For example, we discussed in Chapter 4 how experienced MacDraw users fail to use the full range of keyboard extensions to manipulate objects. Learning by onion-peeling may restrict users to simple or obvious features of the interface.

v.   Since direct manipulation requires actions to be carried out explicitly, there is a danger that task difficulty can be transferred directly to the user. For example, visual programming may enable programmers to link objects together by making pointer selections, but it does not support the task of planning programs. Tasks like programming require some degree of abstract reasoning, whereas direct manipulation is about concrete action. There is a danger that direct manipulation may discourage the planning necessary for complex tasks.

Hypertext, multimedia and virtual reality

Training through exploration is one of the main ideas behind the development of hypertext and multimedia systems, in which a direct manipulation dialogue style is used for browsing through vast databases of interlinked information. The current interest in concepts of hypertext, multimedia and virtual reality can be witnessed through the large sections of conference proceedings such as CHIs 1990–1993 (proceedings of the Conference on Human Factors in Computer Systems) which are devoted to them. To date, the literature has tended to concentrate on technological features of these media, rather than providing evaluations of their psychological characteristics. Rather than simply list here a set of systems and their properties, the interested reader is referred to the proceedings and books on the topic that are mentioned in this chapter. Nonetheless, some of the key properties of these systems are worth outlining here.

Hypertext can be defined as bodies of text-based information which are linked in a non-linear fashion (unlike paper-based text), and which allow the user to browse through the text in any order they choose (for a recent review, see Nielsen, 1990). This contrasts with the traditional book-like representation of information, in which the layout text requires that the user searches forward or backward through series of pages to organize the same set of information under different conceptual categories. With a hypertext system, one is able to create any number of links between related items of information, allowing the user to explore concepts through direct access to related information. The advantages of hypertext over linear text formats are particularly clear in the representation of information which naturally arranges itself in an hierarchical format (e.g., taxonomies, employee hierarchies, etc.).

Hypercard, discussed above in the context of object-oriented programming, is an example of a hypertext system in which presentations of information can be constructed (although some authors refer to it as a 'hypermedia' system, since it has both text and graphics facilities). The idea behind such systems is that the user decides what information to view at any particular stage by pointing the cursor and clicking on text or pictures that are relevant to them. The selection of text or pictures takes the user to related information elsewhere in the database. For example, Figure 8.6 shows a series of cards taken from a Hypercard stack for teaching principles of neuroanatomy. The user is encouraged to explore in increasing detail by clicking on the parts of the brain that they wish to inspect.

Mynatt, Leventhal, Instone, Farhat and Rohlman (1992) compared performance with a hypertext encyclopedia against a paper-based version of the same material. They found that the hypertext encyclopedia elicited superior performance over the paper-based version for questions requiring the location of detailed content (as opposed to titles, etc.). However, this advantage

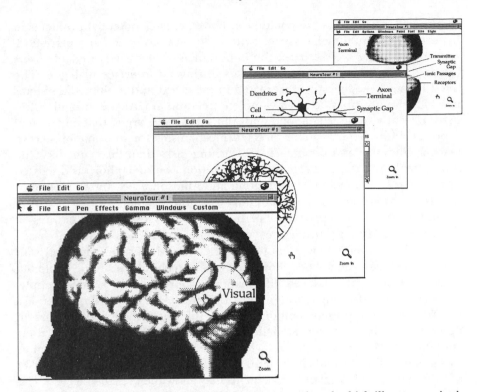

Figure 8.6 Cards taken from NeuroTour, a Hypercard stack which illustrates principles of neuroanatomy. The user browses from one card to the next using the 'zoom' tools, and also accesses hidden text by pointing to regions of interest on the cards.

may have occurred simply because they provided a search tool which allowed users to type in a key word and have the text containing it located for them. This obviously was not available in the paper-based form, but there is no reason why a linear text file (e.g., as used in a word processing system) shouldn't show the same advantages as the hypertext system. In other words, Mynatt *et al.* cannot attribute the advantages of the hypertext encyclopedia to non-linearity, linking, or any of the specific properties which distinguish hypertext from any other electronic text format. It is also worth noting that for most other types of question there was little or no difference between hypertext and paper-based formats. Indeed, the paper-based format was actually superior when the question involved extracting information from non-text sources such as maps, though again this may have been because of a feature of the specific implementation of the encyclopedia rather than because of a problem with the conceptual basis of hypertext.

The provision of enormous amounts of information, together with the removal of the linear representation of books, creates potential problems in hypertext and multimedia systems concerned with navigation and place-finding. In Chapter 5 we discussed the importance of feedback to enable users locate their current position and status during an interface dialogue. The need for such feedback is exacerbated in hypertext and multimedia systems in which the user may be given complete freedom to navigate through information by any route. Hammond and Allinson (1989) argue that a hypertext representation requires more directed mechanisms for information access, learner guidance and other focused learning tasks than linear media. One way of achieving this is through the application of a metaphor (as discussed in Chapter 7). For example, Hammond and Allinson use a 'travel' metaphor to direct students around an exploratory learning system.

Multimedia is the application of the Hypertext concept to mixtures of computer-based text with pictures and sounds from video and audio-based sources. The development of optical and CD-ROM storage combined with enhanced screen capabilities allows this mix of technologies to be exploited in order to present enhanced information resources to users. For example, the British Broadcasting Corporation's DOMESDAY PROJECT consists of a computer-based representation of historical information concerning current demographic information, modelled on the Domesday book, which users (notably school children) can browse to investigate issues such as economic, social and political trends in Britain. By providing easy access to vast bodies of information, it is argued, such systems may allow the user to engage in creative behaviours which previously could not be undertaken with linear information presented in a single medium. Moreover, it has been suggested that hypertext and multimedia systems are particularly promising as learning environments, since the user can determine the route taken through information and the knowledge that is focussed upon.

The concepts of multimedia and direct manipulation are taken a step further in the notion of *virtual reality*. The idea behind virtual reality is to use computer and video-generated graphics to simulate a virtual world for the user to enter, explore and interact with. Typically, a virtual reality system will combine computer-generated graphics with sophisticated input systems such as the data glove (Fisher, 1986) and output systems such as helmet-mounted three-dimensional displays. The user's hand movements can then be sensed and interpreted as part of a computer-generated display which is shown in the user's helmet. Thus there is the potential for users to simulate in real time the manipulation of objects in the virtual display. As with Hypertext and Multimedia, virtual reality is still being developed as a technology, and there has been little or no empirical evaluation of its strengths and weaknesses (for a review of technological developments, see Woolley, 1992). Nevertheless, it is being used increasingly in industrial settings, as well as contexts such as arcade games, where user exploration adds an additional feeling of involvement and excitement.

Practical applications of virtual reality technology include an efficient way of programming robotic arms in manufacturing contexts. The programmer carries out the actions and movements that the robot arm is intended to do whilst wearing a dataglove, and these can be recorded and then played back as movement sequences. Another application is in the simulation of complex and hazardous tasks, such as training micro-surgeons or nuclear plant operatives. If the real world is too dangerous in which to take risks with inexperienced staff, then virtual reality systems offer a potentially safe environment for them to practice and acquire the necessary skills. Finally, virtual reality can be used in the design and testing of complex and expensive systems prior to implementation. For example, the maintenance of aeronautical engines can be simulated before the engine is constructed physically, and any problems can then be fed back to the design team before the engine is constructed.

To date, the psychological concepts behind hypertext, multimedia and virtual reality are largely untested. These are clearly promising technologies which may, like direct manipulation and graphical-user interfaces ten years ago, fundamentally change the kinds of user interface that we will experience in the future. There is always a danger that designers might be carried away by the aesthetic possibilities of mixing computer-generated and video-generated information, and might lose sight of the real objectives of developing training materials. Therefore, it is important to evaluate the concepts that lie behind these systems, such as non-linearity, user-directed exploration, use of alternate representations and real-time simulation.

Training through search

One of the implications of learning through exploration is that the user should be able to acquire knowledge or skills through their own exploration. Training by exploration may well offer a number of advantages over conventional instruction-based methods. For example, Kamouri, Kamouri and Smith (1986) compared the effectiveness of training through exploration with training based on following written procedures. Subjects were trained to control three analogous devices, an alarm clock, a computerized cheque book and a digital radio controller, and then to transfer their skills to an electronic notepad. They found that performance with both the training and transfer devices was better for subjects who trained through exploration. They argue that exploration-based training facilitated analogical transfer between devices, and encouraged subjects to form an abstract representation of the devices which enhanced their ability to control them.

Underlying exploration-based training is the suggestion that *active learning*, in which the learner decides what information or task to focus on at any particular time, is more effective than passive learning. However, users may not always explore a system sufficiently, or may avoid exploring parts of a system in which complex tasks are implemented. For example, Sebrechts and Marsh (1989) describe a study of UNIX command learning, in which both exploration-based learning and model-based instruction (as exemplified by Mayer, 1981) were shown to be partially effective but to have limitations. Exploration-based learning that was entirely unguided led to the acquisition of some command knowledge, but subjects did not explore the system adequately. This result raises doubts about the effectiveness of user-guided search through hypertext training materials that is suggested by authors such as Nielsen (1990). To be successful, exploration-based training needs to be sufficiently compelling that the user will continue to explore the system even when the concepts that the system embodies are complex.

Sebrechts and Marsh also found that subjects who were given a conceptual model of the UNIX system performed better than subjects given no instruction, but only when the instructional materials were present in full throughout training and evaluation phases. In the absence of the instructional materials, the group trained with a conceptual model performed worse than a further group who were trained only with simple descriptive instructions. This result might be interpreted as supporting a 'minimal' approach to instruction (Carroll, 1990) discussed in Chapter 7.

Training to search

So far in this section, we have looked at the design of interfaces for users to search, and we have seen that searching is not always adequate or effective. The effectiveness with which interfaces are searched will be determined by the strategies that users adopt. Therefore, an important area for investigation is the extent to which users can be trained to search effectively.

For example, Green and Gilhooly (1990) investigated the acquisition and training of skills for using MINITAB, a computer-based statistics package. They found that novices could be divided into fast and slow learners. The fast learners employed more exploratory strategies, finding alternative methods for attempting problems and used examples more effectively and with flexibility. Slow learners relied more on a trial and error strategy, and were more repetitive in their use of examples. In a second study, Green and Gilhooly based training on a series of hints which gave either 'effective' strategies that differentiated fast and slow learners in the first study, or 'ineffective' strategies that fast and slow learners had in common. The effective strategies reflect learning by doing, whilst the ineffective strategies were based on learning by thinking (these were judged effective or ineffective to the extent that they differentiated fast and slow learners, but this does not necessarily mean that learning by thinking strategies were ineffective in themselves). Green and Gilhooly found that the training based on the 'effective' hints improved performance significantly, whereas the 'ineffective' hints did not improve performance relative to the no hints group.

The results of Green and Gilhooly illustrate the importance of exploratory strategies in determining the ease with which novice users acquire complex skills. Additionally, they suggest that the effective exploration of interfaces can be enhanced by training in strategies for search. Green and Gilhooly describe the behaviour of their better novices as 'problem finding', that is, a creative search by good learners for commands outside their current repertoire. Clearly there is a danger, with interfaces that rely entirely on user-driven search, that poor learners will fail to explore the interface sufficiently. Thus there is a role for training of skills for interface search. In effect, training such as this is teaching people how to learn, and is likely to become an increasing requirement as the burden of instructional design is placed more on interface designers rather than on trainers.

8.5 Conclusions

In this chapter, a view of skill-as-exploration has been presented. We have seen how expertise involves the application of strategies for gaining information from external displays, thereby reducing both learning requirements and memory loads. The chapter has also examined how interfaces such as object-oriented systems can enhance the creative ability of users and enable them to conduct novel tasks in novel ways. We have also seen how direct manipulation has shifted the focus of task control from the interface to the user. On the positive side, this has reduced the amount of learning required to use an interface. On the negative side, this may have transferred much of the burden of task control from the interface to the user.

There is clearly a complex interaction between the three types of skill (procedures, understanding and exploration) which have been described in Part 3 of this book, and we would not suggest that any interface task is conducted solely with one class of skill in the absence of the others. It is important to recognize that all three types of skill can influence expert performance with an interface. A truly convincing psychological theory of human skill which can predict the occurrence and acquisition of each type of skill and explain the interaction between them is yet to be developed. However, existing theories clearly show the importance of designing interfaces and developing appropriate training regimes to support each type of user skill.

Part 4
Understanding Interface Design

In the final part of the book we consider aspects of the design of interfaces. A central theme of this section is the importance of task analysis in both design and evaluation. Chapter 9 begins the section by laying out the basic processes of design as currently conceived and the problems commonly encountered. Of particular interest is the way in which human interface expertise has been incorporated into the process of design. Chapter 10 looks at the process of evaluation of interfaces and the techniques used to carry it out. Here we compare a number of approaches to evaluation and come to two conclusions. First, evaluation requires expertise to carry out; no methodologies have yet been successfully developed which can be used reliably by non-specialists. Second, surprisingly, the most cost-effective approach to evaluation is also the most simple method. This possibly reflects the predominant role of the evaluator's expertise, rather than methodologies, in the success of evaluation.

CHAPTER 9

Designing interfaces

Overview

In this chapter we examine the processes by which the development of interfaces takes place. First we examine the stages of system development, and we discuss the difficulty of introducing a consideration of the needs of users into existing software development practices. A central theme of this chapter is the use of task analysis as a tool for interface design. Task analysis offers a way of describing the tasks to be accomplished by users of a new device and the conditions under which they can occur, a process which is essential in specifying the features of an interface for the device. We consider some of the additional methods that have been developed to support the design process, such as user-centred system design, checklists and guidelines, and proto-typing.

9.1 Human factors and interface design

Whenever a device, such as a video recorder or a word processor, is to be con-trolled, monitored or read by a human user, there is a need to consider the *human factors* that may affect its use. The term 'human factors' (usually referred to in Europe as *ergonomics*) covers a wide range of issues that affect the use of devices by humans. Apart from the topics covered in previous chap-ters some of these issues focus on physical and environmental characteristics of device use, such as the design of physical layout of equipment to improve posture and comfort and reduce repetitive strain injuries, and control of envi-ronmental conditions to optimize lighting, noise and heat. A further aspect of human factors is the range of psychological issues arising from the process of designing interfaces; these issues are the focus of this chapter.

In a survey of seventy developers of commercial software systems, Myers and Rosson (1992) found that developers spent on average 45% of the time during the design phase of a project, and 50% during the implementation phase, on the user interface. The percentage of code devoted to the user interface was around 50% on average, though this varied enormously (from 1% to 100% depending on the nature of the project). Admittedly, the respondents to the survey were a biased sample (having been approached, in part, through human factors journals), and so probably the amount of time, effort and code spent on the interface by these developers is higher than would be found in many computing companies. Nevertheless, it is clear that the design and implementation of the interface represents a significant, and probably increasing, proportion of the time spent developing computer systems. It follows, then, that the quality of software is becoming increasingly reliant upon the quality of the user interface.

Trying to apply the concepts of human factors to interface design in a commercial or industrial context can be a frustrating business. Some companies are sufficiently forward-looking to recognize the importance of having a human factors input into the design of human-computer interfaces. However, even in these companies, one typically finds that the human factors group is a small specialist team who are brought in to act as consultants at a fairly late stage in the design of new products. They are often asked to 'sort out' the interface, when what is needed is human factors input at the earliest stages in the conception of any product that has an interface with a human user. This chapter examines some of ways in which such an input can be made. It also considers the reasons why this can be a difficult thing to do.

9.2   Stages of system development

Perhaps the most common approach to the development of computer systems and engineering products is an essentially linear set of stages of systems analysis and design. The completion of each stage is assumed to produce the necessary starting conditions for the next stage. This is sometimes known as the waterfall model of system development, the principal stages of which are detailed in Table 9.1 and illustrated in Figure 9.1. The advantage of breaking systems development into discrete stages is that large-scale projects become manageable. It is easier to judge the progress of systems development if it is broken into stages, and time and cost judgements can be made. A description of system development as involving serial and discrete stages is clearly idealized, since it relies on having a completed set of information to pass from one stage to the next. However, one of the key requirements of the waterfall model is that the separation between stages must be clear cut, otherwise the advantages of this approach can be eroded (Curtis, Krasner and Iscoe, 1988). Furthermore, there needs to be some way of deciding when a phase of system development is sufficiently complete so that the next phase can be embarked upon. Herein lies the problem in managing the development of new systems.

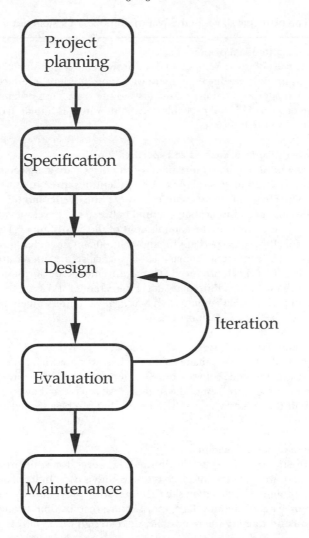

Figure 9.1 A simple 'waterfall' model of system design, showing the separate stages of planning, specification, design, evaluation and maintenance.

It is, in fact, rarely the case that stages in system development succeed each other in an entirely orderly way. In particular, the stage of articulating a *requirements specification* (see Table 9.1) is often incomplete or disorganized. The reasons why a coherent specification is not produced are many, and include: the sheer quantity of information involved; the inherent difficulty of knowing what is required; the difficulty of describing known requirements in a clear and unambiguous way; and problems caused by several people being involved in the process. A poor requirements specification can in turn cause

Table 9.1   The principal stages in the Waterfall model of system development.

---

1. Project conception and planning
Projects may consist of the development of entirely new devices, updating existing devices or re-equipping existing equipment with more modern interface technology. At this stage, managerial decisions are made as to the scale, cost, organization and priority of the project. This will probably determine whether human factors specialists are used to develop the interface.

2. Requirements capture, analysis and specification
This stage is one in which the requirements of the device are gathered into a requirements specification. Easterbrook (1991) identifies three roles which the specification plays: first, it forms a contract between the clients and the developers; second, it is the main communication channel between the developers and the clients; and third, it establishes the commitment of the contributors. Thus, the requirements specification is perhaps the most significant document produced during a project. Unfortunately, human-factors specialists are often introduced into a project only after it has been produced. The requirements specification often serves as a document against which third-party developers tender. Pressure to accept the cheapest tender creates what amounts to a bias against tenders which include potentially expensive human-factors work.

3. Conceptual and detailed design
During this stage, decisions are made about the best way in which to get the required functionality. In enlightened software development companies, the interface will be designed in tandem with the rest of the software. All too often, however, the interface has been seen in the past as a kind of 'front end' that can be attached at a later stage to the software.

4. Implementation and evaluation
By this stage the interface is essentially finished. However, for some projects this might be the first time that the interface is tested with a user. Indeed, occasionally it is only when equipment is installed on site that operating procedures are worked out. When problems arise during this stage, they require rectification through documentation and training where possible, or re-design where necessary.

5. Maintenance
Maintenance in the context of interface development usually involves fixing technical bugs. By this stage it is usually too late for any redesign of the interface to take place. However, problems in using the interface can make it difficult to distinguish between bugs in the functioning of the device and bugs in the interface itself.

---

major problems for the designer. If the specification is too loose this may not give sufficient guidance to the designer; if it is too tight the design may be constrained and feasible options excluded.

In order to accommodate some of these difficulties, it is common that a project will involve some degree of iteration between stages. For example, if during a phase of conceptual design it emerges that the performance requirements for an interface are insufficiently specified, then a further phase of requirements specification must take place. At a later stage, the outcome of evaluations may also lead to some redesign. Clearly, the more iteration that must take place, the less efficient and more costly is the design process.

A number of structured design methodologies have been developed to maintain the segmentation and orderliness of a systems development process (e.g., SSADM – Ashworth and Goodland, 1990; Information Engineering – Olle, Haggelstein, MacDonald, Rolland, Sol, Assche and Verrijn-Stuart, 1988). Structured approaches aid the understanding of, and navigation through, the complexities of design projects by providing methods such as *functional decomposition*, in which a large design problem is broken down into a number of smaller and functionally discrete sub-problems. Methodologies employ a variety of documentation techniques and notations to organize the process of functional decomposition, such as decision trees and tables, data flow and structure diagrams, and 'structured English'. These techniques are intended to reduce the burden of handling design information and enable greater control of the development process.

Whilst methodologies such as SSADM offer mechanisms for structuring the design of computer-based systems, they have a number of drawbacks and limitations, of which the most important are:

1. Complexity – These approaches require substantial amounts of documentation, and also skill on the part of the designer. For example, SSADM has seven stages, three graphical notations and a vast range of technical or semi-technical jargon associated with it;
2. Inflexibility – As a result of their emphasis on functional decomposition, methodologies tend to be product-oriented and may not pay sufficient attention to the nature of the environment in which systems will be implemented. This can be particularly problematic for interface designers, where the concern is primarily with the match between the system and the user, rather than the functionality of the system itself;
3. Inhibition of creative design – Strict adherence to a structured approach can have the unwanted by-product of encouraging designers to be conservative. If development of a system is guided only by managerial targets, then unambitious system design based on 'yesterday's philosophy' may occur (Parnas and Clements, 1986; Vidgen and Hepworth, 1990).

Even if these problems with methodologies are alleviated, it is still difficult to incorporate human factors into a stage model of systems development. There are two main reasons why this is the case:

1. The distinction between system and user requirements – The stage approach does not differentiate between functional requirements of the

system and information requirements of the user. Whilst information and functional requirements are obviously related, they are not the same thing. Information requirements describe the information that the user needs to carry out the functional requirements for which the interface is designed, but they must do so in a way which is psychologically feasible. While methodologies support the specification of functional requirements, they do not always consider how they will be expressed at the interface in a way which results in good design from the user's point of view. The examples of screen design discussed in Chapter 3, such as the help screen shown in Figure 3.9, illustrate the poor designs that can result from considering only functional requirements.

2. The separation between requirements specification and design phases: In the waterfall model, requirements specification is a discrete stage that ideally is completed before system design starts. It is natural, therefore, to make the requirements specification as detailed as possible. However, a consequence of increasing the detail is that the specification can become too rigid to allow the consideration of alternative technical solutions. For example, a specification that requires a user to *type in* a command choice may be unnecessarily restrictive compared with one which requires the user to *indicate* a choice. The development of alternative and innovative technical options is a mainstay of design, and has featured strongly in the recent development of interface technology. This is illustrated by the development of graphical user interfaces, which followed from the technological innovation of pointer-driven input systems such as the mouse. The result has been to alleviate many of the human factors problems that previously faced users (e.g., the need to learn large numbers of text-based commands). A managerial separation between specification and design stages decreases the ease with which such innovative design can proceed.

9.3   Task analysis as a design tool

Despite the difficulties of getting a proper consideration of human factors into the design process, research into human-computer interaction has identified many of the properties that good interfaces should have, and the preceding chapters have outlined many of the design options for supporting users that have emerged from this research. What is needed, then, is an effective method for deciding which design options are appropriate for a new interface. We believe that the most useful method with which to approach this problem is *task analysis*. Task analysis, as the name suggests, entails a thorough and detailed description of the complete set of objectives that a user may be required to achieve through an interface, as well as the conditions under which tasks can and should be undertaken. Task analysis can be used for both the design of new interfaces and the evaluation of existing interfaces. Task analysis supersedes other approaches to interface design because of its potential for organising all aspects of specification, design and evaluation. Indeed, we would go so far as to suggest that some form of task analysis should be a prerequisite for the development of any device that has an interface with a human user.

Different design objectives naturally lend themselves to different approaches to task analysis. Since the reason for analysing tasks are many, there are many techniques which could be called task analysis. For example, in a recent book on the subject, 25 different techniques are described (Kirwan and Ainsworth, 1992). The term 'task analysis' has been used for a wide range of purposes: to describe the user skills that must be trained; to ensure safety in complex processes; to undertake assessments of cost-effectiveness of activities; and to judge attitudes towards a system. An equally wide range of methods for conducting task analysis have been suggested, including verbal protocol analysis, critical incident analysis, time and motion studies, and the use of questionnaires. In this chapter, however, we are interested in the role of task analysis in developing requirements specifications that encourage designers to choose from a range of options the optimum design features for a novel interface.

One way of approaching the analysis of complex tasks for which an interface may be used is to see them as an arrangement of less complex sub-tasks, which themselves may comprise a number of elements. We have adopted Hierarchical Task Analysis (HTA), described by Annett, Duncan, Stammers and Gray (1971), as an example of the kind of task analysis which is being used increasingly in the design and evaluation of interfaces. For example, HTA has been shown to be particularly well suited to describing process control tasks (Shepherd, 1985). This is not to imply a unique importance for HTA over other methods, although we believe it to be a method of considerable flexibility and simplicity (for descriptions of different methods of task analysis, see Clegg, Warr, Green, Monk, Kemp, Allison and Lansdale, 1988; Diaper, 1989; Kirwan and Ainsworth, 1992). Figure 9.2 shows an HTA analysis of the task of operating an overhead projector, a task which has been broken down (at least partly) into its constituent elements (from Shepherd, 1986).

Figure 9.2 illustrates how HTA describes tasks in terms of the goals that users are required to attain. Analytic methods concerned with representing tasks in this way invariably produce hierarchical descriptions of users' goals. If a broad statement of the user's goals is too general to be of use to designers, then the analyst can re-describe the goal in terms of a set of sub-ordinate goals and a *plan* which governs the conditions for carrying out the constituent sub-goals. A plan contains information concerning the sequence in which sub-tasks are carried out and the conditions that must exist before they are undertaken. Examination of these sub-goals highlights the feedback that the user must monitor to determine how to regulate actions.

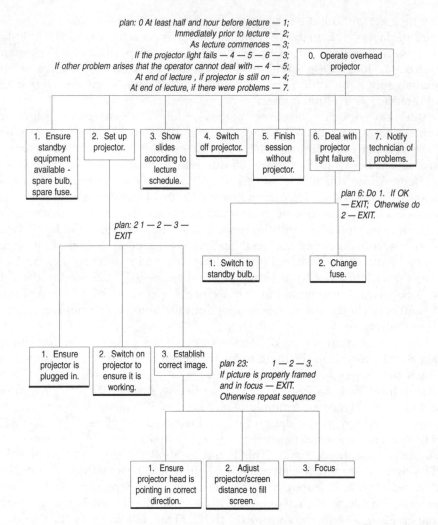

Figure 9.2 An Hierarchical Task Analysis of operating an overhead projector (Shepherd, 1985).

To describe fully the use of task analysis in interface design, we need to consider in more detail the reasons why designers might wish to analyse tasks. There are three main reasons for carrying out task analysis: first, to understand the logical structure of complex tasks; second, to identify the information that users require to carry out a set of tasks; and third, to understand the activities that people actually undertake in using an interface, and the conditions that affect their performance. These are discussed in the sections that follow.

Identifying relationships between task elements

A particular advantage of HTA for the design of new interfaces comes from the hierarchical description of a device in terms of the user's goals rather than in terms of the tasks currently carried out by users with an existing interface. By preserving an aspect of task description in terms of goals, the designer is able to see the goal that has to be attained, not simply the users' present behaviours that have to be supported. This enables an astute designer to offer novel solutions to the problems they encounter, since there are always a number of technological options which can satisfy a user goal.

Hierarchical representations of users' tasks are of particular value in considering design options because of what has been described as the 'paradox of change' (Carey, Stammers and Astley, 1989). This refers to the problem that the methods employed in achieving an objective often change with the tools provided to do it. Changing the interface for a new system can in turn change peoples' objectives and their behaviour. The paradox is that the automation of tasks often leads to a redesign of the task itself. In such a climate, making recommendations about changes to an interface without understanding the effects of such changes on users' tasks is attempting to hit a moving target.

For example, administrative practices alter as written communications are replaced by electronic mail systems: documents such as circulars (in which a single document is read and ticked-off by individuals as it travels around a circulation list) are replaced by electronic mail messages. It is also becoming common to store such information on publicly-accessible sources. Comments and proposed amendments to documents may be logged on a collective resource, rather than an individual receiving comments and collating them. The result is not merely a change in procedure. The public debate which follows, and the visible record of it, change both the kinds of remarks people are likely to make and the way in which the instigator of a debate sets it up in the first place.

Changes in administrative procedure that occur through the introduction of new technology have become the focus of research into computer-supported cooperative work (CSCW). We do not deal with CSCW in detail in this book, partly because the kinds of interfaces that might best be provided to support cooperative work have yet to be determined. Also, we remain to be convinced that designing interfaces to support cooperative work requires anything more than the task analysis and design methods we suggest should be used in the development of interfaces for individuals (though, this is not to say that interfaces for individual users will necessarily be sufficient to support multiple users, as the above arguments imply). Nevertheless, there are many researchers, both computer scientists and psychologists, who believe CSCW should be distinct topic of human-computer interaction research (for reviews see Greenberg, 1991; Bowers and Benford, 1992).

The benefit of task analysis as a tool for interface designers lies in its ability to separate out as far as possible the technology-independent and technology-specific elements of a task. Hierarchical representations of tasks usually start with the most abstract, broad elements at the top of the hierarchy and become increasingly specific and concrete with depth. This feature is particularly valuable because the higher levels of analysis tend to be technology-independent and are designed to represent the goal structures of the task independently of existing task constraints (which generally appear lower in the hierarchy). The ability to stand outside the existing solutions is an important factor in the objective design of interfaces. It serves to free us from the tyranny of seeing tasks in terms of existing practices, which may largely have been determined by the constraints of an old or defunct technology.

As well as assisting the designer in making decisions about design options, task analysis enables an evaluation to be made of the mapping between system functionality and user actions. This is possible because the end-product of task analysis represents the various components of a task in a coherent representation. For example, HTA uses a diagram of the task hierarchy to show the inherent structure and relations between tasks (illustrated in Figure 9.2), as well as a tabular representation to allow significant comments about each task to be recorded. Inconsistencies of action between functions which share similar objectives, and pseudo-consistencies between functions (where similarities of method do not necessarily result in the same objective) can be revealed by examination of these hierarchies and tables. Knowledge of these consistencies can inform better interface design, training methods, and documentation.

A difficulty in using HTA as a method for evaluating consistency is that the hierarchical breakdown of tasks is described in natural language. The clarity of this description depends, therefore, on the skills and linguistic usage of the person who conducts the analysis. This problem can be overcome by the provision of formal or restricted notations with which to re-describe the products of a task analysis. In Chapter 4, we described how TAG (Task-Action Grammers: Payne and Green, 1986) offers a restricted notation which was devised explicitly for the purpose of assessing the degree of consistency that exists in interface dialogues. As we argued in before, TAG is not in itself a method of task analysis. Instead, TAG is best used as a notation for re-describing the products of a task analysis in terms of the actions that are required to undertake the users' tasks.

Identifying information needs

In specifying a new system, task analysis serves to indicate the elements of functionality that a specification must cover. One role of task analysis, therefore, is to scope the functionality of an intended product, an important process in the production of a requirements specification. More than this, however, a good task analysis will show how the functionality maps onto users' tasks, and will indicate the conditions that are necessary for these tasks to be accomplished. In particular, it will highlight the *information requirements* of the user, that is, the information that users need at any one time to perform a specific task.

For example, a common project in the domain of process control is to upgrade operations rooms by replacing older control displays, whose indicators are attached by pneumatic links to the valves, pipes and tanks of the plant, with modern computer-controlled VDU displays. Pneumatically-controlled displays tend to be panels, in which dials, alarms and level indicators show the status of the plant equipment. These are located on the panel in approximately the same spatial layout as the plant itself. In the process of converting from a panel to a VDU display, some subtle but nonetheless important information is lost which relates the position of the display instrumentation to the equipment that it controls or monitors. Furthermore, the amount of information that can be viewed at any one time is highly restricted by the size of a VDU screen. Thus, whilst technological upgrading may maintain the same functionality of a display, it frequently changes the way in which information is made available to the operator. In order to understand the implications of changes such as this, the designer must be able to identify the information needs of operators for each task that is carried out using a control display.

An obvious approach for a designer in replacing old interface technology is to use the same kind of information representations in the new system that were used in the old display (e.g., using an electronic rather than pneumatically-controlled floating indicator to show the level in a tank). Indeed, it is often argued that it is better to maintain the old representation in a new system because the operators will be able to transfer their skills from the old to the new and will therefore require less re-training. However, as the example of replacing panel displays with VDUs indicates, technological changes can have hidden effects on the information needed by operators to carry out their tasks. We know of at least one case, in a waste re-processing plant, where the old displays had to be re-installed after they had been replaced because operators were unable to carry out many of the plant monitoring tasks using the new VDU displays. Since the plant managers insisted that the VDU technology be used, the number of control-room staff was nearly doubled to meet the extra (and unnecessary) demands that were placed on operators of monitoring two sets of control displays. In other similar cases the response has been to simulate a panel display by having rows of VDU screens stacked one on top of the other. This is, however, a very costly solution to the problems created by the introduction of new technology.

Although the replacement of old interface technologies with new ones may create difficulties that the designer must recognize, it can also create opportunities for enhancing users' performance. If the information needs for every task that a user carries out are identified through task analysis, then it is possible to devise novel and creative ways of using new technologies to provide the right information at the right time for each task. The replacement of panels with VDUs need not necessarily impair the performance of process-control operators. Indeed, Shepherd and Ormerod (1991) have argued that VDU technology actually provides an opportunity for designers to restructure the display in such a way that the right information can be displayed in the optimum format on a VDU screen for a specific operational task.

For example, the task of monitoring an information display to anticipate a target level in, say, a pipe-flow does not necessitate continuous monitoring. Instead, the operator can be alerted by a pre-alarm to attend to a simple digital display of the current pipe-flow just before the target is reached. On the other hand, the task of monitoring to detect deviations in pipe-flow necessitates continuous monitoring. The skilled operator will make adjustments as soon as a there is a trend away from normal levels and before the level goes out of tolerance. This task may best be served by a cumulative-trend display, in which the relationship between the current flow and maximum and minimum tolerance thresholds is indicated. These alternatives are illustrated in Figure 9.3.

To enable designers of process control interfaces to identify the information needs of operators, Shepherd and Ormerod (1991; see also Shepherd, 1993) have developed SGTs (Sub-Goal Templates) as a methodology for converting an HTA into an information-requirements specification. SGTs are essentially a notation for re-describing a task analysis in terms of the common sub-goals that are set by operators in conducting their tasks. At the broadest level, sub-goals can be classified as intentions to perform different kinds of monitoring, communication, action and diagnosis activities. Each of these sub-goals can be further specified until the information needs for each class of sub-goal can be identified. For example, a monitoring activity can be undertaken in order to anticipate a change, to inspect a rate of change, or to detect deviations. To carry out each type of monitoring sub-goals may require the same set of information (e.g., the rate at which a liquid is flowing through a pipe), but the way in which that information is represented can be optimized for each type (e.g., through representation as either cumulative trends, maximum-minimum thresholds or alarms). Thus, in the same way that TAG can be used to re-describe the users' tasks in terms of the actions that are undertaken to execute the tasks, SGTs can be used to re-describe users' tasks in terms of the information that users need in order to perform each task.

Identifying user activities

One of the common sources of confusion in the use of task analysis is whether it produces a logical representation of a task, or whether it describes

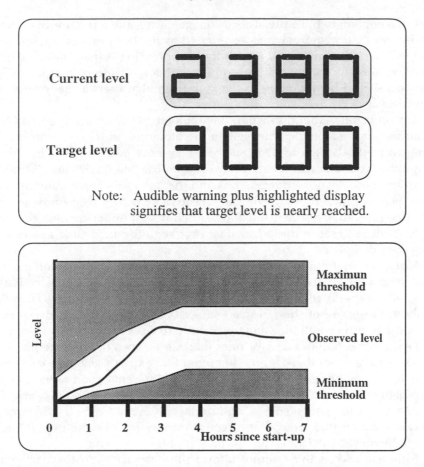

Figure 9.3 Two alternative displays for the same data (a pipe-flow level). The first display requires a pre-alarm to notify the operator just before a required level is reached. The second display uses trend information to enable the operator to make adjustments to the flow in sufficient time during a process to avoid crossing a maximum or minimum tolerance threshold.

the psychological processes behind what users actually do when carrying out a task. Different protagonists of task analysis differ in their emphasis in this respect. For example, HTA as applied by Shepherd (1989) is strongly aimed at *avoiding* a psychological specification of users' performance, whereas as a nominally similar method, TAKD (Diaper and Johnson, 1989), explicitly seeks it. The proliferation and overlap of methods of task analysis has led to terminological confusion in the area. We choose to describe approaches which claim to provide a psychological specification of the skills and knowledge that users have to perform tasks as being methods of *cognitive modelling* (e.g., TAKD). Approaches which make no assumptions about the nature of users' knowledge, but instead focus on specifying the structure of complex tasks and their requirements, we refer to as methods of *task analysis* (e.g., HTA). The key distinction between these approaches is that, whilst cognitive analysis must by its very nature propose a theory of users' knowledge, task analysis should be neutral as to the knowledge that users actually possess to perform tasks.

Unfortunately, the distinction between cognitive modelling and task analysis is not maintained throughout the literature, with many writers referring to methods such as TAKD as being *cognitive task analysis*. The idea of cognitive task analysis seems to us to confuse an important distinction between a user's knowledge of a task and the task itself. Since different users may possess different knowledge about the same task, it seems only sensible for designers to focus their initial design efforts on understanding the tasks, hence our preference for task analysis over cognitive modelling as a tool for interface designers.

Under our definitions, the outcome of a task analysis of an interface does not necessarily represent something of psychological significance. Rather task analysis, and re-description notations such as TAG and SGTs, reflect only the structure of the tasks and sub-tasks that the system is designed to support and the conditions under which these tasks can occur. For example, a TAG analysis outlines a set of rules that can be used to implement a set of tasks as actions, but there is no assumption that expert users have knowledge which corresponds to these rules. In practice, different expert users may conceptualize the functionality of the interface in quite different ways. The result is that the consistency of an interface as evaluated by a TAG analysis may not match the perception of consistency as seen by the user (Grudin, 1989; Reisner, 1990; Draper and Jordan, 1993).

Although task analysis cannot inform the designer as to the nature and extent of users' skills, it can nonetheless provide information about the ways in which users perform tasks. Specifically, a task analysis which is based on observational studies of users can indicate the extent and frequency with which each sub-task is performed. This information can then be used to design an optimised interface. For example, the organization of items within a hierarchical menu system should reflect the order, precedence and frequency with which different tasks are carried out. Commands for primary, important or frequent tasks can then be placed as menu items at the top of the hierarchy (i.e., within the main menu), whilst secondary, minor or infrequent commands can be placed further down the hierarchy (i.e., as choices offered after a selection from the main menu is made). Another example of interface design based on task analysis of task order, precedence and frequency is offered by Card, Moran and Newell (1983), who describe the design of optimized key pad systems based on an analysis of the keystrokes necessary to implement tasks through the user interface.

An evaluation of task analysis

Whilst most of the derivatives of task analysis are complex, as our worked examples of TAG given in Chapter 4 illustrate, we would argue that HTA offers a relatively simple approach to explicating the structure of complex interface tasks. An important practical feature of HTA is that the task is described in operational terms which can be approved by the client and understood by the contractor; it is not necessary to have a specialist human factors training to understand a task description using this method. Therefore management prerogative is preserved and effective communication between the contributing parties is maintained.

However, even HTA requires some degree of skill and experience on the part of the analyst. Every method of task analysis has a subjective element, and often the methods reflect the biases and domains of individual researchers. It is perhaps symptomatic of the subjectivity of different methods of task analysis that different techniques tend to be related to local research cultures. TAKD, for example, has only been documented in use by a collection of workers originating from the same laboratory. The use of HTA and TAG can similarly be traced to common roots. This may reflect the fact that the success of these techniques is based upon the experience that individuals have in applying them. Whether these techniques can be passed on without support from experienced practitioners (for example, by reading accounts of how to do task analysis, as exemplified by Diaper, 1989) is not clear. We see little evidence that successful task analysts actually develop their skills this way. More likely, analysts who already have experience in the area come to specific techniques such as TAG or HTA as a way of formalizing their approach to evaluation. In Chapter 10, we return to the same issue in the context of evaluation, rather than design.

Paradoxically, the intention of task analysis is *not* to produce *the* definitive description of a task. It is conceivable, for example, that two analysts using HTA might produce two different but internally consistent descriptions of a task. In essence, subjectivity is not damaging when task analysis is seen as a *process* rather than providing a *product*. If task analysis is seen as a process, then regardless of the inherent subjectivity of the method, simply undertaking the activity of trying to specify and re-describe tasks is a very informative process for the interface designer. Where a task analysis is leading to a product, such as in providing the basis for a requirements specification, it is important that it is verified through further observation and discussion. Redescription notations such as TAG and SGTs provide a way of enhancing the rigour of task analyses, since the application of these methods highlights aberrant features of the analysis, in the form of inconsistencies or unclassifiable sub-goals.

9.4   Choosing between design options

As the contents of Chapters 2 to 8 demonstrate, there have been considerable advances in research into human-computer interaction over the past twenty years, and interface design alternatives continue to proliferate. The problem for the designer is to choose the option which will best serve the users of a new interface. In the previous section we argued that task analysis is a prerequisite for interface design. However, whilst task analysis is a useful tool for identifying the requirements of an interface, it does not *in itself* provide the design options that might best meet these requirements. Ideally, every interface designer would be an expert in human-computer interaction and psychology, and would be able to apply their experience and skills in the design of interface features. Thus, one approach might be to equip designers with human factors expertise through training (and books such as this one offer a limited but, we hope, useful basis on which to develop such expertise). In practice it is clearly not possible to re-train every computer scientist or engineer who might build a device which has an interface with a human user. Therefore, what is needed is an effective way of allowing people other than specialists in human-computer interaction and psychology to apply human factors expertise. In this section we look at three broad approaches to this:

1. Supplying designers with advice about optimising interface features. This approach may take many forms, but is seen most commonly in sets of guidelines, standards, company style-guides and checklists.
2. Encouraging design from the user's perspective: This is often described as *user-centred system design* (UCSD). UCSD is not so much a method as a philosophy about the focus of system design. Essentially, the intention is to make the user an explicit, if not primary, focus of the designer's efforts. We use the term to describe, not only the general philosophy, but also methodologies which include the user as a participant in the design process and also approaches which take a broader view of the social and organizational context in which interfaces are used.
3. Iterative design of interfaces: As we described earlier in this Chapter, systems development rarely proceeds in an entirely linear fashion. Instead, components of the system tend to be specified, designed and evaluated, and then re-specified and re-designed according to any emerging flaws or possible improvements. This process can be turned into an advantage for the designer who adopts *prototyping* as an approach to interface design. Simple prototypes of interfaces can be constructed and then evaluated before the designer must commit the system to a specific interface.

These three approaches overlap considerably. For example, prototyping forms an essential component of most participative methods of interface design. Also, all three approaches tend to be used in parallel by experienced interface designers. Nevertheless, each approach has its own strengths and weaknesses, which we evaluate in the sections that follow.

Guidelines, standards, style-guides and checklists

A common approach to delivering human-factors knowledge in a seemingly applicable form is by producing guidelines (e.g., Smith and Mosier, 1984; Cole, Lansdale and Christie, 1985) or 'toolkits' for design and evaluation (e.g., Galer, Harker and Ziegler, 1992). These offer, in the form of rules or checklists of varying degrees of specificity, lists of important issues and advice about good and bad features for interfaces which the designer can work through in the course of designing an interface. Ideally, a well-conceived set of guidelines will require little or no specialist knowledge and be relatively concise, yet offer advice on the most significant design features. Within large companies there are often company style-guides which contribute to system development. For example, British Telecom, IBM and Apple all offer style guides.

This general approach is particularly attractive because, firstly, it aims to be directly useful to design and therefore promises immediate benefits compared with experimental research or usability evaluation; secondly, the general aim is to deliver expertise in a form which can be applied by others, thereby making the human-computer interaction expertise available to as many people as possible and delivering it reasonably cheaply compared with employing a human-computer interaction consultant directly. A third benefit of these products, at least for those who produce them, is that they are tangible: they are clear evidence to industries of the worth of an human-computer interaction or human factors team. Such 'products' are therefore a means by which human-computer interaction teams can demonstrate their own productivity; something which otherwise is rather hard to determine.

Increasingly, the expertise behind guidelines has been turned on its head and used to produce *standards*. Ideally, these statements represent mandatory criteria against which interfaces are judged (e.g., see Stewart, 1990; Brooke, Bevan, Brigham, Harker and Youmans, 1990) where procuring or legislative bodies have felt the need to ensure good design and/or to protect the user public. Such standards are indications of requirements designed to ensure that designs do not fall outside acceptable limits. They are, not surprisingly, very hard to formulate in the area of interface design because so little of human behaviour is understood well enough or sufficiently predictable to lend themselves to standardization. Consequently, such standards are often so loosely formulated as to mean almost anything. This is exemplified by the inclusion of requirements on software usability given in a recent European Community directive, and shown in Table 9.2. Some standards apply in areas where much more is understood about the relationship between the

Table 9.2   The requirements for enhanced usability as specified in a recent European Community directive.

"... In designing, selecting, commissioning and modifying software, and in designing tasks using display screen equipment, the employer shall take into account the following principles:

(a) software must be suitable for the task;

(b) software must be easy to use and, where appropriate, adaptable to the operator's level of knowledge or experience; ...

(c) systems must provide feedback to workers on their performance;

(d   systems must display information in a format and at a pace which are adapted to operators;

(e) the principles of software ergonomics must be applied, in particular to human data processing".

(90/270/EEC, 1990, articles 4 and 5, annex 3)

---

technology and human behaviour. An example of this might be in the area of VDT standards, reflecting a considerable understanding of the parameters and functions of the human visual system. However, even here, standards have caused concern because they derived from technology which does not apply to other display technologies.

We have two main concerns about the use of approaches which offer guidelines and other sources of advice to designers. The first is that their use in unskilled hands can still lead to poor design. Worse, a poor design might be seen to be legitimized by reference to checklists and guidelines which are developed by human-computer interaction specialists. A checklist typically asks the user to consider some rather general points and is relatively insensitive to the context in which the application must be made. Thus, what is suitable for a particular type of task in one context may not be suitable for another context (Maguire, 1982). We argue that, where checklists are used to help make design decisions, they must be sensitive to the contexts in which tasks are taking place. It may be appropriate to employ a checklist approach to make practical decisions, but only after the full implications of the task and the influence of context are understood. Thus, it would be sensible to use a checklist approach only after careful task analysis has been undertaken.

The second concern about the utility of guidelines, style-guides and checklists is their long term value. As interface technologies change, so the design options and difficulties also change. For example, guidelines describing the best colours for presenting text on VDUs have changed as the technology has advanced. The same is, of course, true for any method which communicates the current state of the art in human factors and human-computer inter-action. Nevertheless, the problem is particularly acute for guidelines and checklists because they tend to be delivered in the form of rules. This is problematic because people have a nasty habit of treating design rules, irrespective of their validity, as if they were cast in tablets of stone!

User-centred system design (UCSD)

The design process commonly focuses upon technical and economic factors but consideration also needs to be given to the demands of the user and the context in which a device will be used. A technical specialist will look at the system requirements in terms of factors such as reliability, capacity, speed, cost, compatibility with existing equipment. However, addressing human factors issues will produce a different set of criteria for assessing design require-ments. Eason (1988) classifies these criteria under four broad areas: functionality; usability (see Chapter 10 for a further discussion of the con-cept of usability); user acceptability; and organizational acceptability. These are detailed in Table 9.3. These criteria reflect the fact that a technical sys-tem has to serve not only the specific tasks for which it is introduced, but also has to fit the wider, pre-existing context in which it has to operate. However, the accelerating rate of technical developments reduces the time available to address these issues in the deployment of new systems.

Table 9.3   Eason's (1988) criteria for assessing design requirements

---

Functionality: The technical specification must cover the functions the system will have to be able to perform in order that it can support the required range of organizational tasks.

Usability: The system must offer its functionality in such a way that the planned users will be able to master and exploit it without undue strain on their capacities and skills.

User Acceptability: The system must offer its services in a way which its users will perceive, as a minimum, as not threatening aspects of their work they hold to be important, and, ideally, they will perceive it as positively facilitating goals they wish to pursue.

Organizational Acceptability: The organization at large has goals, policies and structures and the system must not only serve immediate task needs but must not impede other aspects of organizational functioning. Ideally it will serve as a vehicle to promote wider organizational goals; as a minimum it must provide an 'organizational match'.

UCSD is an explicit attempt to address these criteria through changing the focus of system development to include a consideration of the user. UCSD can be seen to occur in three main ways: first, in attempts by human-computer interaction researchers to change the general focus of systems development to include the user and not solely the function of the system; second, to promote the idea that the user should become actively involved in the process of systems development; and third, to expand the notion of the 'user' to consider the social and organizational context in which the system will be used.

The first use, and the one adopted in the book *User-centred system design* (Norman and Draper, 1986), effectively demands a complete understanding of a user's goals and information requirements. A user-centred approach sets up the question "What information will the user need to operate this device?" as the central theme motivating its design. The designer's job is then to deliver that information to the user in a useful form. When an explanation of how to use a device is obvious from the structure and layout of the device itself, the device offers an *affordance* to the user (a concept introduced in Chapters 4 and 5 of this book). In other words, the design supports the information requirements of users in a natural and intuitively obvious way.

A second meaning of the term 'user-centred system design' is taken to imply user involvement in the design of interfaces. A stage approach to systems development does not make explicit provision for the needs or opinions of users to be fed into the development of the requirements specification. Participative methods require an explicit input into requirements specifications from end-users. In participative approaches users are expected to contribute to, and gain from, the design of the system, and this should increase the likelihood of its success. For example, ETHICS (Effective Technical and Human Implementation of Computer-based Systems: Mumford, 1979) is a participative methodology for systems design.

Whilst participative approaches appear attractive, they tend not to detail the nature of user input. If anything, the end-user is simply introduced as part of a cycle of iterative design and evaluation. The end-user does not generate the information requirements that designers receive, but merely evaluates the prototypical designs which emerge from functional requirements specifications. This can be problematic, because users tend to show a natural preference for interfaces which are familiar to them or which are superficially more attractive. Thus potentially effective, but nonetheless unfamiliar, novel design solutions can be rejected by users even though they offer a better interface than more familiar solutions.

The third use of the term broadens consideration of the word 'user', in that the user is perceived as being more than the individual who conducts a dialogue with an interface. Eason's (1988) criteria of usability described in Table 9.3 offer a *socio-technical* view of user-centred design, which include effects of the social, organizational and political contexts in which an interface might be implemented. The ETHICS methodology also encompasses the socio-technical view, that for a system to be effective, the technology must fit closely with the social and organizational factors that determine the context in which it will be used.

Prototyping

Prototyping entails the construction and evaluation of mock-up interfaces during the process of system development. The aim of prototyping is to enable an immediate evaluation to be made of the potential advantages and disadvantages of major design decisions at an early stage before commitments are made that cannot be broken. Prototyping plays an important role in user-centred system design, since it enables both the designer and the user to comment on the proposed interface before the system has been designed in its final form, which should reduce the likelihood of user dissatisfaction (Harker, 1987). Gladden (1982) suggests that, in providing physical objects, prototyping can convey more information for designers than written specifications. There is some empirical evidence which suggests advantages for a prototyping approach. For example, Stephens and Bates (1990) found that using a prototyping approach in a poorly specified design task led to more effective progress than using structured design methods. Prototyping has been enabled by the increased speed and availability of software tools, in particular by fourth-generation programming systems. Myers and Rosson (1992) classify such tools under four headings, illustrated in Table 9.4. For example, Figure 9.4 shows the toolbox displays from Microsoft's Visual Basic, a prototyping tool which provides access to the Windows interface tools.

Whilst prototyping is clearly an attractive option for interface designers, as witnessed by the extensive use of prototyping tools reported by Myers and Rosson, it also has potential pitfalls. First, some designers who use prototyping methods concern themselves only with the user interface and do not address the fundamental problems of system analysis. In other words, they are simply making poor systems superficially palatable to users (an attractive prototype has led to the acceptance of many a poorly-conceived computer system). Second, constructing prototypes can be a time-consuming business, especially if a number of iterations are required to get it right. Third, there is often a reluctance to throw existing prototypes away. This happens partly as a consequence of the time spent in developing prototypes, and possibly because the presence of a prototype stifles the designer's ability to think of alternatives.

A consequence of these difficulties is that early prototypes are often developed further and integrated into the final product. This may seem quite reasonable, given the lines of code that a software developer might have to generate for a prototype. However, it can also mean that poor prototypes can exert a strong influence on the final design. As lecturers, we encounter an

Table 9.4 Examples of Myers & Rosson's (1992) classification of prototyping systems

---

1. **Window managers:** these are software packages that divide the display of the development environment into discrete areas for different development tasks and contexts (e.g., Microsoft's Windows environment for IBM compatible personal computers).

2. **Toolkits:** collections of interface components such as menus, buttons and scroll bars, which a programmer can harness as pre-defined routines. Many operating systems make such toolkits available (e.g., the Apple Macintosh toolbox). Hypercard, for example, offers an easy way to access some dialogue components and build mock-ups of interfaces.

3. **Interface builders:** These are graphics-based tools, usually with a direct manipulation dialogue style, which can be used to create and place interface components on a display. For example, LPAs MacProlog™ Dialog Editor can be used to create dialogues for Prolog programs. The designer can place dialogue objects on the screen and then resize them, and the Dialog Editor then provides the Prolog code that is necessary to produce these dialogue screens when the developer's program is run.

4. **User Interface Management Systems:** These are basically more sophisticated interface builders, which support not only the creation of dialogue screens but also the linking of these with the program that the dialogues are intended to interact with.

---

analogous situation in teaching students how to write essays. We sometimes think it better to persuade students *not* to write out draft versions of essays, since they tend to use draft copies as an excuse for postponing the hard work of planning their essays, believing that they can re-structure the essay once they can see its shape in draft form. Unfortunately, more often than not, the only changes made between draft and final copy are improvements in hand-writing, spelling and (occasionally) grammar.

In fact, this problem is general to all software development. Important design decisions that are left until late are difficult to enact when they involve radical changes because of the investment of effort that the design process has already built in. Equally, poor decisions that are made early can exert their influence throughout the course of design and may be impossible to unpick later. Brooks (1975) observed that software developers often assume that problems in design can be dealt with at a late stage in the development process, because a programming language is a symbolic rather than a physical medium. This is because lines of a program can be altered at any stage without necessarily disrupting the rest of the code. Therefore, software design is seen as rather more flexible than domains like architectural design, where one cannot alter the foundations once the roof is on. However, as

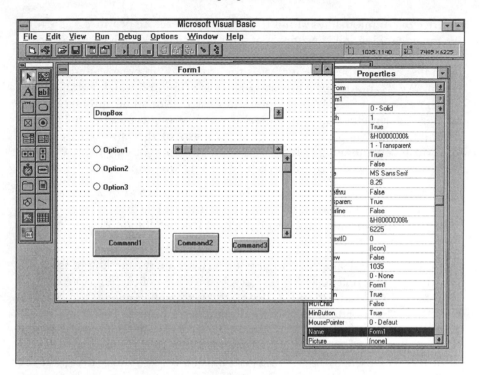

Figure 9.4 Visual Basic, a prototyping tool for interface design (the form shows some examples of dialogue features which can be added using the tools shown in the icon bar on the left).

Brooks points out, the consequence of this is that decisions are often not made sufficiently early, so that by the time deficiencies in the design are uncovered it is too costly to modify the software.

9.5   Conclusions

Given the range of different methods for supporting interface design that are available, which one(s) should the designer employ? The answer, unfortunately, is not simple. Choice of method will depend on the complexity of the design task, the cost of device construction, the human-factors experience of the designer, the timescale of the project, and a multitude of other factors.

We have argued in this chapter that, of all the methods available to interface designers, task analysis is probably the most significant. When carried out carefully, it gives the opportunity to generate specifications which are sensitive to commonalities within the task, and therefore provides opportunities for enhancing the consistency of an interface. Task analysis also decreases the likelihood that specifications will be influenced by existing technologies, and therefore increases the likelihood of creative and effective design. We do not claim that task analysis is a method which automatically delivers insights which lead to optimum design. Nor would protagonists of any particular method such as HTA make such strong claims for their particular approach (Shepherd, 1985). The strength of the approach seems to come from encouraging a form of lateral thinking (arguably an essential element in creative design) which is based upon two propositions:

(a) There is a logical distinction between the *objectives* of a task and the *artefacts* and *procedures* used to achieve those objectives;

(b) There is an inter-dependence between the evolution of tasks and the evolution of interface artefacts.

The strength of task analysis is to provide a richer understanding of the complexity of the problem at hand. Carroll and his colleagues (Carroll and Campbell, 1989; Carroll and Kellogg, 1989; Carroll, Kellogg and Rosson, 1990) have described the process of design as reflecting a task-artefact *cycle*. Interface design can be seen as an iterative process in which the development of new artefacts (such as computer-based systems) encourages a new set of tasks for users to carry out. These in turn require the re-design of the artefacts, and so on. The introduction of some new device or new functionality in an existing device makes possible, often unexpectedly, some new way of doing things or reaching objectives not previously possible. The proliferation of unexpected uses for spreadsheets provides a good example of this (Hewitt, 1989). Originally designed for financial analysis, they are rapidly developing as educational tools, research simulators, databases, and often supplant other programming systems.

CHAPTER 10

Evaluating Interfaces

Overview

Evaluation is an important activity in the design cycle. In recent years, it has become popular to develop practical methods which can be applied to the evaluation of interfaces. The term 'tools', as in 'software evaluation tools', or 'usability assessment tools', has often been used to describe these methods. Like the battery of tests a medical consultant has available to diagnose a patient, each of the different approaches to evaluation has its own strengths and weaknesses. They also differ in their cost and the technical competence required to carry them out. The purpose of this final chapter is to show that evaluation requires expertise and is not a simple procedure which can be safely applied without some understanding of the issues discussed in the rest of this book.

10.1 Evaluating the usability of interfaces

In the previous chapters, we have used the words *usability* and *evaluation* rather freely. Human-computer interaction is an area in which terminology can be confusing because the same words can be used to mean different things by different people (see, for example, Young's, 1983, discussion of the term 'mental model'). Therefore, we start by considering what the words evaluation and usability mean and how they should be used:

> **Evaluate**: To work out the value of; to reckon up, ascertain the amount of; to express in terms of the known. (*Oxford English Dictionary*).

Suppose you were asked to evaluate a chair. What would you measure – cost? size? How would you put a value on comfort? Would we require an office chair to be as comfortable as, say, an armchair or does that fact that we use them for different things make comfort a poor dimension to evaluate? Suppose that we find that comfortable chairs cost more money. How do we trade off comfort against cost? Further trade-offs come to mind. For example one car might have comfortable seats, but after a drive down the motorway gives us backache, whereas another seems rigid and less obviously comfortable but does not give us backache. The difference seems to occur because of the posture each seat imposes. If we are to evaluate seats, how do we include considerations of posture into the equation? These questions illustrate two fundamental issues in evaluation: First, how do we put a value upon the dimensions we evaluate? Second, how do we decide what dimensions to evaluate in the first place?

Usable: That may or can be used. (*Oxford English Dictionary*)

The term 'usability' has been coined to describe the set of properties that should be evaluated in an interface, or to describe the quality a good interface should possess. However, a useful definition of usability proves to be elusive. Consider those given in Table 10.1. Although in an everyday sense these are reasonable definitions, they are somewhat unhelpful. In particular, they give no indication of the processes required for carrying out an evaluation. In truth, most definitions of usability are circular, because usability is defined by the things one measures to judge the ability to do things: Usability is the ability to use!

The circularity of definitions is not lost on those who write about usability. Chapanis (1991), for example, openly recognizes that the words *easily* and *effectively* are inexact, and looks at several different operational definitions, based on error scores, execution times, learning time, subjective user ratings, and so on. Implicit in this approach is the need for a variety of methods for, and approaches to, usability evaluation. The problem, in the absence of a

Table 10.1   Some definitions of usability.

"[Usability] concerns the extent to which an end-user is able to carry out required tasks successfully, and without difficulty, using the computer application system." Ravden and Johnson (1989, p. 9).

"Usability is presented as a concept which can limit the degree to which a user can realize the potential utility of a computer system." Eason (1984, p. 133).

"The usability of a computer is measured by how easily and how effectively the computer can be used by a specific set of users, given particular kinds of support, to carry out a fixed set of tasks, in a defined set of environments . . ." (Chapanis, 1991, p. 3).

". . . the capability in human functional terms to be used easily and effectively by the specified range of users, given specified training and user support, to fulfil the specified range of tasks, within the specified range of environmental scenarios." Shackel (1991, p. 24).

---

formal definition of usability, is to decide which methods to use, and how the results of these different approaches are ultimately to be integrated. In this sense, evaluation is more a craft than an exact science.

To avoid the problem of circularity, some definitions take a more empirical line in which usability is defined in terms of objective and measurable criteria. This is the philosophy behind Shackel's (1991) definition given in Table 10.1. This definition is aimed at defining what aspects of user performance are important, and the context in which performance is set, and then

Table 10.2 A worked example of setting usability goals via the definition of usability, in this case an electronic mail system. (From Shackel, 1991.)

To achieve the target usability, tasks 3–8 shall be done:

Effectiveness

at better than 2 seconds for each task with no more than one error per 50 attempts by 90% of the target range of managers, secretaries and professionals in any office situation which complies with the offices, shops and factories acts and which does not seriously contravene the workstation and environment specifications recommended in the V.D.T. Manual (Cakir, Hart and Stewart, 1980) or in Designing Systems for People (Damodaran, Simpson and Wilson, 1980)

Learnability

either within 2 hours of starting use via learning with the manual and help system or within 1 hour of starting use via the training course

based upon the 'getting started' manual to be provided or upon the 2 hour training course to be prepared

within half-hour relearning time for users who have (1) completed the learning defined above, (2) done all the tasks once per week in half-hour sessions for 3 further weeks, and (3) then not done the tasks for an interval of 4 weeks

Flexibilty

with the task flexibility requirement only to apply to tasks 1, 2 and 9, but with the environment flexibility requirement to apply to tasks 3–8 that user performance shall not deteriorate more than 10% in warm conditions up to 95 degrees F and 95% R.H.

Attitude

and with attitude questionnaire results on 5-point scales ('very good' to 'very bad') to be at least 80% in 'good' or better and only 2% below 'neutral',

so that questionnaire results also give at least 90% 'yes' answers to the question "Imagine you must use this system 5 times per day every day – would you be satisfied to continue using it, and if not please comment why not?"

---

setting for each aspect a performance criterion which has to be met in this respect. Shackel (1991) gives examples as to how this is done, as illustrated in the example shown in Table 10.2, which describes usability goals for an electronic mail package.

The declaration of rigorous and (more importantly) useful definitions of usability is sometimes referred to as *usability engineering* (e.g., Whiteside, Bennett and Holtzblatt, 1988), reflecting the emphasis upon designing or 'engineering' interfaces to a given requirements specification. An example of a usability engineering approach is given in Table 10.3, which illustrates the cases of worst, acceptable, best and current values of the time taken to carry out an installation task. There are a number of benefits in adopting a usability engineering approach. For many situations it may be possible to define specific criteria and to achieve them in design. In some situations, such as safety critical systems, they may be a regulated requirement. Also, the very act of producing usability specifications, and getting them agreed by a design and evaluation team is a means by which design can be focused around a set of agreed issues. It can therefore act as a coordinating influence in design.

Usability engineering, however, has one major drawback: the setting and achievement of benchmarks and performance criteria does not really assess *usability*. The problem is not the use of benchmarks (although we should recognize that it is not always be possible to design an interface that reaches the required performance criteria), but of deciding which are important criteria to set. In practice, interfaces designed in this way will still need an evaluation to check whether unforeseen problems have arisen in design.

Evaluation is a process of measuring interfaces and *usability* reflects a vague notion of quality. In practice, both are rhetorical, rather than truly technical, concepts. The real purpose of evaluation in most cases is not measurement for measurement's sake, but to inform a *critique* of an interface with a view to making design alterations. Critiques are more than simple quantifications. Different kinds of measures are useful for supporting a particular point of view about an interface, but they do not, in themselves, determine that view. In this sense, therefore, *evaluation* as commonly used in human-computer interaction is possibly a misnomer, since the key outcome is often qualitative rather than quantitative.

Table 10.3 A sample row from a usability specification (workstation installation). (From Whiteside, Bennett and Holtzblatt, in Helander (Ed.), 1988, p. 795).

Attribute	Measuring Concept	Measuring Method	Worst Case	Planned Level	Best Case	Now Level
Installability	Installation task	Time to install	1 day with media	1 hour without media	10 min. with media	Many can't install

Equally, 'usability' could be seen as a misnomer. It is an anti-concept more easily identified when it is absent than when it is present (see, for example, Eason's quote in Table 10.1). Since problems in the use of interfaces can come from a wide range of interacting factors, this makes an attempt to define *usability* futile. More realistically, usability is simply a catch-all term which serves to remind us of the need to design interfaces from a human point of view.

10.2   Different reasons for evaluation

There are many different ways of putting a value on something, depending upon what you choose to measure and how you choose to measure it. As advertising agencies, politicians, and opinion pollsters have known for some time, this is not an objective process. Consequently, the measure of an evaluation lies not just in what result was found, but how that result was arrived at, and why the evaluation was carried out in the first place. There are a number of reasons why evaluations of interfaces are carried out. Although the distinction between them is not hard and fast, it is useful to list them under four headings:

Formative evaluation

Formative evaluation is also known as iterative or developmental evaluation. The basic idea is that 'cheap and cheerful' testing can serve to iron-out obvious problems before the design process has gone so far as to make changes impractical. Allowing this feedback loop to operate fast enough in the design process requires prototyping tools which maximize the speed and minimise the cost of mocking-up a trial interface. Prototyping as a method of formative evaluation can be seen as a design exercise, and its effect upon the design process is not clearly understood (see Chapter 9). In this Chapter, however, we are concerned with how evaluation might be carried out, rather than how its results impact upon design.

Summative evaluation

This is an holistic approach in which a fully-formed interface is evaluated across a range of factors. Summative evaluations are often carried out by organisations such as The Consumer's Association with a view to informing consumers as to which products to buy. The basic problem in this type of evaluation is that the motivations of those who want to examine the results of an evaluation may differ. Consequently, a wide range of factors are often evaluated to inform the reader about different aspects of the product. This can lead to two problems in summative evaluation. First, the evaluator has to make decisions about what factors are relevant, which can be difficult: and second, the communication of an evaluation across a range of dimensions can be very complex (see Roberts and Moran, 1983, for an example of an evaluation of word-processors across many dimensions). This means that such evaluations are hard to understand. Second, they are highly context-specific: If the circumstances change, for example, if a new range of word processors uses speech recognition or freehand input devices, a different range of evaluative factors might be appropriate.

Trouble-shooting

Rather like doctors attempting to diagnose a medical condition and prepare treatment on the basis of a set of presenting symptoms, consultants in human-computer interaction are often called into situations in which a number of difficulties have arisen. Their role is usually to diagnose the root cause(s) of the problem with a view to making recommendations as to how the problem can be resolved. Evaluation in this case is aimed at understanding the factors underlying the visible problems. It must also offer practical and effective strategies to remedy the situation. Like summative evaluation, trouble-shooting can be complex because of the potentially wide range of factors involved. Further, the cause of a particular set of problems may be related only indirectly to the observed symptoms, and consultants need to consider a wide spectrum of issues in human-computer interaction issues.

Meeting benchmarks

Occasionally, designers must ensure that users operate at or above specific performance criteria. For example, controllers of nuclear reactor systems can allow for no error and may have to make fast or immediate responses to certain types of error condition. The setting of such operational criteria, followed by evaluation to check they are met, is central to the usability engineering approach outlined in the previous section. However, evaluation of this kind is rarely seen, for two reasons. First, interfaces whose conditions of use are so critical that absolute performance benchmarks are needed are often found in areas where confidentiality is important, such as in military applications or safety-conscious process control plants. Second, for most interfaces it is difficult to identify which aspects of performance to specify as important, and the level of performance which is required. Consequently, although it is an attractive idea that good interface design might be ensured by the setting of standards against which design can be tested, in practice it is not common.

10.3   Evaluation methods

We have already commented upon the choice that evaluators have in deciding what attributes of an interface to measure. More generally, this reflects
the fact that the *process* of evaluation – what you do, and how and when you
do it – is worryingly unstructured. This inevitably leads to a proliferation of
approaches to evaluation, and also gives rise to differing performance
between evaluators. For example, Hammond, Hinton, Barnard, Maclean,
Long and Whitefield (1984) found that different evaluators of the same
system overlapped only slightly in their appraisal of it.

Evaluators may overlook important aspects of an interface, or may over-
emphasize certain aspects at the expense of others. This leads to a desire to
develop standardized, structured approaches to avoid these problems. Also,
evaluations can be costly and long-winded. This contrasts badly with the fast
turnaround often required in design projects. Therefore, methods by which
evaluators can focus reliably upon key design issues in an efficient and cost-
effective way are at a premium.

In this section we describe five common approaches to evaluation, discussing their main features, advantages and disadvantages. These are
summarized in Table 10.4.

Usability testing

Usability testing is based upon the principle of trialling prototypes and capturing data about the system and user performance which can then be analysed. One of the difficulties in this approach is that 'performance' is an
all-embracing term. Usability testers are presented with the problem of
deciding which data to collect and analyse. We discuss below seven types of
data which are commonly collected by evaluators in usability tests.

Usability testing is hard and usually expensive. It requires judgement as to
the application of the methods, and requires skill and experience in setting
up the trials and analysing the data. For the outcome of these trials to have
validity, it is important that the users tested are representative of the intended
user population and that the tasks they are required to carry out are realistic
and, furthermore, are treated realistically by the users. This can be difficult if
the intended user population is small and specialised, or if the tasks concerned are hard to simulate, as in hazardous situations. Such testing often
requires well-developed interfaces. Prototyping tools are becoming increasingly quick at producing acceptable mock-up interfaces, but it is still sometimes difficult to fit user testing into the design cycle until the interface is
well developed. By this time, of course, the opportunities for change are
reduced and the value of evaluation therefore limited.

Table 10.4 Some of the potential advantages and disadvantages of five different approaches to usability evaluation (adapted and extended from Jeffries *et al.*, 1991).

Approach	advantages	Disadvantages
Usability Testing	Identifies serious and recurring problems; avoids low-priority problems; some degree of objectivity	Requires user-interface expertise; high cost; may require large sample of users; misses consistency problems
Guidelines and Standards	Identifies general and recurring problems; can be used by non-specialists; applicable at all stages of design	Misses some severe problems; can be mis-applied to defend bad designs or even to create worse designs
Formal Modelling	Provides quantitative analysis of interface properties; can give unexpected insights; some degree of objectivity	Extremely complex and requires expertise; tends to focus on only one (usually high-level) interface dimension
Walkthrough Approaches	Less complex than formal modelling; provide structure for evaluations	Require expertise; can be long-winded; use complex notations; subjective; miss important problems
Heuristic Evaluation	Identifies many problems; identifies more serious problems; low cost; can predict further evaluation needs	Requires some degree of expertise; requires several evaluators; some degree of subjectivity

i) Throughput

Throughput is a measure of productivity. It might include the number of pages proof-read, menus navigated, problems solved or forms dispatched. When two interfaces supporting the same task are being compared, throughput measures provide a reasonable way of differentiating them. Problems arise in such comparisons when the task is not exactly the same in each case, such that simple measures of throughput can become misleading. An example of this is the comparison of hypertext vs. paper-based encyclopedia carried out by Mynatt *et al.* (1992) which we discussed in Chapter 8. Explanations for the productivity of the hypertext tool over its paper-based equivalent were based upon the presumed advantages of hypertext (which stem mainly from the non-linear structure of text). However, the hypertext system provided a search tool which was not available to the paper-based system, but was in principle implementable in a linear system such as a word processor. Consequently, Mynatt *et al.* cannot attribute the advantages of the hypertext system to non-linearity, or any of the specific properties which distinguish hypertext from other text formats because they failed to produce comparable tasks. The same argument applies to any form of comparative evaluation, be it based upon throughput or any of the other measures discussed below.

ii) Execution time

The converse of measuring the number of specified tasks carried out in a given time is to measure the time taken to carry out a given operation. The usual assumption is that longer performance times reflect poorer design: They are usually associated with greater difficulty, reduced throughput, and user frustration. For example, Resnick and Virzi (1992) plotted selection times on two different menu designs as a function of the position of the target in the menu. The results are shown in Figure 10.1. The longer time to select a given target in one design was taken to demonstrate its inferiority. The same method is applied widely to the evaluation of many different interfaces. For example, Palmiter and Elkerton (1991) investigated animated demonstrations as a means of training users of direct manipulation interfaces. They used speed of performance as a measure of the relative benefits of animated demonstrations over procedural textual instructions. Their results showed that the animated demonstrations group performed more quickly after training, but the text group was quicker after a delay of one week, suggesting that the animated demonstrations produced less durable learning than text-based learning.

Differences in the times taken to carry out a task with different interface designs are often minuscule, and hard to establish statistically without many observations from a large number of subjects. However, small differences can add up to a great deal of time (i.e., money) and user frustration if the applications and screens are commonly used by large numbers of users. For example, Fisher and Tan (1989) argue for the use of highlighted data on displays. Given the vast number of person-hours spent viewing some screens of frequently-used applications, speed improvements of a fraction of a second resulting from evaluation and redesign of data displays are likely to be cost effective. Such an argument is often used to justify devoting substantial amounts of evaluation effort to applications which are aimed at large markets.

iii) Accuracy

Accuracy usually refers to performance in which the precision, rather than the correctness, of inputs is at stake. For example, in a drawing package, a

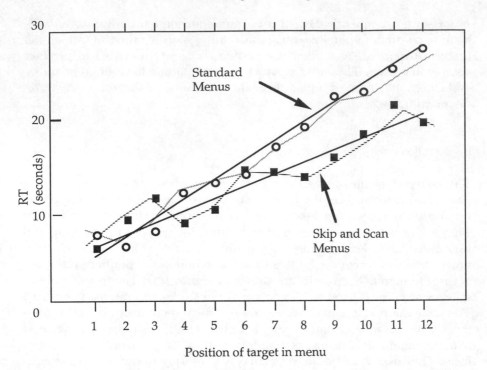

Figure 10.1 Plots of selection times for two designs of telephone menus as a function of the position of the target item in the menu list. The Skip and Scan style reduces the need for the users to wait through long recitations by allowing them the facility of control over the recordings, equivalent to the opportunities for visual scan in screen-based menus. The plots show a clear speed advantage for skip-and-scan menus as the target is placed further down the list, as might be expected (from Resnick and Virzi, 1992).

question might arise about the users' ability to align or place items in the right place. Inaccuracies do not invalidate the input but reduce its quality, which is distinct from the error of misspelling a keyed-in command word or choosing the wrong command. For example, Jellinek and Card (1990) compared variable-gain mouse input with ordinary mouse input. Here a measure of comparison was the users' ability to control the cursor accurately and quickly with one of two types of mouse.

iv) Errors

When user actions are clearly distinguishable as either correct or incorrect, performance can be evaluated in terms of *errors*. Errors can be used to infer difficulties of interface use deriving from a mismatch of interface design and the user's psychological processes. The problem is that errors, when they can be identified, can arise from a range of causes. Simple mechanical slips or inaccuracies may activate process A when process B was intended (as with the accidental selection of adjacent items on a pull-down menu); memory failures might cause wrong inputs of data; and poor screen design can result in misinterpretation of the state of the system. The analysis given in Chapter 5 is an attempt to categorize errors in terms of their causes and to argue, where appropriate, for design changes when frequencies of errors with an identifiable cause reach 'unacceptable' levels.

Formal, large-scale, evaluations may generate the need to undertake detailed analyses of error frequencies, sequential dependencies (how likely it is that error A follows circumstance B), severity, time taken to correct them, and many other statistics. If a detailed comparison of two interfaces is being undertaken, it may be important to know how likely certain errors are and what they are likely to cost in terms of time and money. For most evaluation procedures understanding *how* these errors occur is enough, sometimes in combination with other sources of data such as user's reports, to enable the evaluator to see the potential for certain types of error in the interface.

v) Subjective measures

Many usability tests incorporate an assessment of users' subjective responses, often in the form of questionnaires or rating scales. They might record attitudes to a system; preferences between alternatives; and estimations of how easy things are to do. It seems obvious in user testing that the user's remarks and attitudes are useful information as part of a battery of testing. For example, if users are slower using interface A than interface B, then a marked preference for B would seem to support the conclusion that B was better.

There are two problems to the use of subjective measures in evaluation. First, what people say in response to subjective rating methods is highly sensitive to the way in which they are asked. In practice it is very difficult to design questionnaire and rating methods which are not open to biased interpretation. The second problem with subjective measures is that they can be misleading: what the users *say* they like and what is ultimately the best design solution are not always the same thing. For example, in the evaluation of animated demonstrations of computer-based procedures, Palmiter and Elkerton (1991) found that such demonstrations, although preferred by the subjects, were not always as effective in training as other methods. As in other areas, 'the user knows best' is often found wanting as a guideline for interface design.

The problem of interpreting subjective assessments is also related to the time-course of the evaluation. Only if the study is carried out over a long time are the users' responses likely to reflect those of a representative user population. If the study time is too short, two types of error are likely. First, if the new technology is innovative and superficially attractive, novelty effects can lead to positive responses which might wear off in the long term if the interface fails to deliver the functionality required. On the other hand, negative responses sometimes arise when a new interface appears to be hard to use, unfamiliar, or restrictive in some way. Very often, such assessments are graded upwards as the user population discovers that the interface is more effective than previously thought.

vi) Memory load

Many difficulties in the use of an interface emanate from the user being required to remember more than is reasonable. For example, in an invoicing system studied by Lansdale and Newman (1991), users could be required to move into a different program to find an account number which they needed to do a different job. The time taken to suspend the first job, find the necessary data in the second database, close the database and then restart their original task was in the order of minutes. Given the complexity of the order number (an alphanumeric string of 13 characters), and the distractions of negotiating between programs, the observed delay of about two minutes was too long for most users. Tell-tale symptoms of memory loads might be users making notes on paper of the information they need later, users breaking the task into smaller or less efficient components, and errors resulting from typing in the wrong information.

Assessing the nature of a memory load is not straightforward for two reasons. First, errors due to memory failures are often difficult to distinguish from other causes. Second, there are no simple guidelines as to when to expect memory load difficulties: How much people can be expected to remember depends partly upon what they already know. For example, the digits 29483-35423-59235 might seem a lot to remember and more than most people can cope with, which according to Miller (1956) is normally no more than seven or so unrelated items. However, one might find them easy to remember if they are all telephone numbers of friends. Essentially, the more grouping of information into meaningful units that occurs, the more information will be remembered. Conversely, the more arbitrary information is, the harder it is to learn and remember. An example of the organization of this is shown in Table 10.5.

Note that there are two possible sources of memory overload. It can come from the user being asked to hold too much data in mind, as illustrated in Table 10.5. It can also come from the procedural complexity of the task that the user is being asked to carry out. Most objectives have subgoals, contingencies and interdependencies: 'To do X, do Y followed by Z unless B is true, in which case do Q.' Here the user's working memory can become

Table 10.5 An example of evaluating memory loads, showing an interaction between task, expertise and memory load.

Suppose users have to remember the following keyed sequence (21 characters in all):

LHR/JFK-BA702-228-tue

This presents a memory load that is considerably larger than the maximum of 5–7 items suggested in most guidelines (e.g. Smith and Mosier, 1984), and it supposedly exceeds the capacity of working memory. It would not be surprising, therefore, to see users make errors in reproducing this sequence. A possible set of errors might be as follows:

LH*W*/JFK-B*S*702-228-tu*r*

However, errors made in performing this task would not necessarily be as random as those shown above. In the context of booking an airline ticket, the letters and numbers in the sequence are related to each other and so the units that must be remembered are bigger than individual keystrokes. The sequence shown above actually means the following (where brackets enclose the sequence to be typed):

London Heathrow (LHR) to (/) Kennedy Airport (JFK) by flight (-BA702) at price (-£228) on Tuesday (-tue)

Consequently, errors tend to reflect a larger unit of analysis than individual characters. An example of errors we might expect to see would be:

LHR/*LGA*-BA702-228-*wed*

suggesting that the user forgot the destination and the day of departure. Skilled users would probably learn that BA702 is a scheduled flight between London and New York, and would therefore be less to make an error about the flight destination. This leaves only the price and day to remember, which is a significantly smaller memory load than the 21 character sequence at first appeared.

---

over-loaded in such a way as to make it hard to keep track of where they are, which parts of the task are done, and which are not. The procedural complexity of interface use and its effect upon working memory is one of the properties of interfaces investigated by the use of formal modelling methods in evaluation which we discuss below.

For most evaluations of memory load, measurement techniques are usually *ad hoc* and rarely formal. Questions such as "Are experienced users prone to errors from memory requirements which are too great?" (taken from a checklist in Clegg, Warr, Green, Monk, Kemp, Allison and Lansdale, 1988) are often sufficient to reveal whether memory loads are an issue. If users are not able to say clearly that they have difficulty remembering or keeping track of something, it is unlikely that there is a substantial memory load problem. Apart from asking this simple question, therefore, the skill required in memory-load evaluations to assess what the nature of the memory load really is, once it has been established that a problem of this nature really exists.

vii) Video tapes and system logs

Many systems allow for the capture of data which records the user's inputs to the system and their timing. A number of laboratories also use video recording of users' behaviour in usability trials (e.g., Good, 1985). In the case of UNIX, the availability of an event logging system has also spawned a number of studies aimed at understanding UNIX expertise by analysis of these logs (e.g., Bradford, Murray and Carey, 1990). Similar approaches have been used to study electronic mail usage (Mackay, Malone, Crowston, Rao, Rosenblitt and Card, 1989).

Although it is easy to capture these data, it is more difficult to analyse than discrete measures of performance. A major problem is to categorize sequences of behaviour, be they keystrokes or movements on a video. For example, a user spending a long time navigating through a number of information screens might be looking for a specific piece of information (possibly indicating a problem) or just browsing (which is not necessarily a problem). Also, analysis is immensely time consuming. Evaluators often spend hours poring over relatively short sequences of video recordings or analysing the tabulated record of all the users' interactions in a specific session. This arises because important and informative sequences of behaviour are often embedded in large amounts of irrelevant or insignificant data.

One approach to this problem is to attempt to automate it. For example, Siochi and Hix (1991) describe a study in which 17,086 command lines were taken over a period of three months from 17 users and were analysed for *maximal repeating patterns*, that is, repeated actions representing symptoms of interesting usability problems. There are, of course, problems with such approaches. For example, the system log needs to be in a form amenable to such analysis, which is not always the case. Nevertheless, the development of analytic tools complements the repertoire of usability testing and may well be useful in certain circumstances.

In our opinion, analyses of system logs and video tapes only rarely provide insights into the use of interfaces which could not also have been found from other, less labour-intensive approaches. Where such approaches do have a value is in establishing and communicating issues which have been identified by other evaluation methods. For example, on the basis of a relatively simple assessment of memory load, one might assert that the software requires the user to remember too much. By concentrating upon those errors which can be ascribed to the forgetting of information, statistics from logged data can be more compelling than an informal analysis. Such tools only make sense in the hands of a skilled practitioner who is able to use their judgement and experience to collect and analyse the data. Otherwise the danger is that considerable time and money is spent collecting data, 99% of which is of no value and/or unanalysable.

Guidelines, checklists, and standards

A checklist, such as those presented in Ravden and Johnson (1989) or Clegg *et al.* (1988), is a sequence of questions noting the attributes of an interface. Guidelines (e.g., Smith and Mosier, 1984) are effectively a checklist by which designers and evaluators can work through a range of design recommendations, systematically evaluating certain aspects of the interface against the guideline concerned. Standards are guidelines of varying specificity which become statutory requirements and which manufacturers are expected to meet.

Potentially, there are advantages to these approaches, both in design and evaluation. First, they are seen as *transferable*: human factors specialists can (in principle) formulate guidelines and checklists which can then be used by non-specialist designers and evaluators. As such, they would be flexible and might reduce the demands upon the skills of interface specialists. Second, these methods are cheap, accessible sources of human factors input which can, theoretically at least, be applied early in the design and evaluation cycle. Finally, such approaches can be structured in such a way as to draw a distinction between flexible heuristics, whose purpose is to orient evaluators quickly and cheaply towards important issues, and checklists, which allow for thorough examination of detailed issues (see Clegg *et al.*, 1988, for an example of this distinction in practice).

In practice, however, these approaches seem very limited in value. Recent studies by Tetzlaff and Schwartz (1991) and Thovtrup and Nielsen (1991) emphasize how hard guidelines and standards are to use. Further studies which compare the use of guidelines and standards with other evaluation approaches suggest that they are less effective in identifying serious usability problems, are hard to use, and expensive in their need for supplementary expertise (these are discussed further in the next section). It is conceivable that checklists such as those provided by Clegg et al. (1988) and Ravden and Johnson (1989) may sometimes be useful in *orienting* the experienced evaluator towards certain issues they may not have considered, but as we shall see, there may be easier ways of doing this.

Formal modelling as an approach to evaluation

Research in formal modelling has the aim of developing theories about interaction at interfaces which can be used to structure more objective techniques for the evaluation process. Formal models attempt to define the interactive properties of the interface and the cognitive skills required to use them. Formal models have been associated with the following advantages (following Olson and Olson, 1990):

1. They can limit designs to psychologically plausible alternatives. For example, they can avoid overloading working memory;
2. They can be used in deciding between two alternative designs for an interface;
3. They can estimate performance times for given tasks;
4. They can contribute to the development of training regimes and inform designers as to the most efficient methods;
5. They can indicate 'hot spots' of interface use where errors or delays can be expected, hence directing attention towards significant performance issues.

A number of approaches to formal modelling have arisen in recent years, foremost amongst which is the GOMS model (Card, Moran and Newell, 1983). Other prominent approaches, some of which differ only in subtle ways from GOMS, include TAG (discussed at length in Chapter 4), CCT (Kieras and Polson, 1985), and newer approaches such as PUMs (Young, Green and Simon, 1989). These acronyms, and the models they stand for, are described below.

Formal modelling methods fall into three overlapping categories:

a) They are exploratory tools in research. The attempt to develop formal models of interface performance (of both user and system) reveals interesting insights and develops methodology further. To some extent, all the models described below fall into this category.
b) They represent theories of cognition and interfaces. At present it seems optimistic to think that the models developed will qualify as scientific theories. Indeed, Carroll and Kellogg (1989) question whether current psychological research is in fact capable of such theories.
c) They are useful disciplines for the evaluation of interfaces. Later sections of this chapter will argue that this role in itself does not justify their complexity.

i) GOMS (Goals, Operators, Methods and Selection rules)

The GOMS model of Card, Moran and Newell (1983) is a method for classifying the skills necessary to use an interface in terms of Goals (i.e., user intentions), Operators (i.e., individual decisions and response options that are available to the user), Methods (i.e., the combinations of Operator applications that can achieve a goal) and Selection rules (i.e., rules for choosing between alternative methods which achieve the same goal). The purpose of GOMS is to describe idealized, error-free behaviour in terms of these four concepts and to provide predictions for the time taken to execute given tasks. Using known times for specific operations, GOMS can be used to evaluate interfaces insofar as it predicts execution times for given tasks with different interfaces without the need to run usability trials.

Since 1983, GOMS has reappeared regularly in numerous guises as researchers and practitioners attempt to extend the use of the technique to practical problems. Thus a number of variations have been developed to accommodate specific difficulties of the GOMS approach, such as CPM-GOMS (Critical Path Methods – Gray, John and Atwood, 1992) and NGOMSL (Natural GOMS Language – Kieras, 1988).

The attraction of GOMS is the potential for a formal method to deliver an objective technique for evaluating and designing interfaces. The other formal methods discussed below, arguably all related to GOMS, share this aim. It is too early to say whether GOMS and its variants will eventually meet this aim. However, in its original form, a number of difficulties have been identified in the use of GOMS which will need to be addressed if it is to be of general use in evaluation:

a) Technology dependence: a GOMS analysis requires a specification of an interface before it can be formulated. GOMS analyses are therefore context-specific and do not access the functionality of an interface;

b) Errors: GOMS does not model errors, particularly conceptual errors or those arising from individual differences, expertise or fatigue. Further, GOMS assumes expert (i.e. error-free) performance and has nothing to say about the components of skill, pre-expert performance, or forgetting of skills;

c) Inflexibility of method: To formulate predictions about the time taken to carry out tasks with an interface assumes a single method for carrying out a given task, with the components being executed serially. This is necessary to carry out the task-action mapping which is an essential part of the GOMS philosophy. In using interfaces, however, a number of studies have shown that users often use a variety of methods, sometimes in parallel, and not always the most efficient ones (e.g., Olson and Nilson, 1988);

d) Restriction of tasks: Since GOMS is strongly proceduralised by its analytic technique, it applies best to simple tasks. Recent approaches have attempted to widen GOMS to more complex tasks by using critical path analysis (e.g., Gray, John & Atwood, 1992). This essentially applies a GOMS analysis to the sequence of activities in the use of an interface which takes the longest time. By identifying the slowest sequence of activities it targets those which will impose the greatest limits on user performance.

The broadening of GOMS to less restrictive tasks is problematic. For example, in attempting to apply GOMS to 'browsing' tasks, Peck and John (1992) argue that browsing, which they initially studied because of its assumed complexity was, in fact simple and predictable, and easily subsumed into a GOMS formulation. Such a finding is counter-intuitive, and suggests that applying a GOMS analysis to complex activities might lead to interesting insights about user behaviour. However, this is not so much an extension of the GOMS technique as a re-definition of the task to which it is applied as being simpler than previously envisaged.

ii) TAG (Task-Action Grammars)

TAG (Payne and Green, 1986) is discussed in considerable detail in Chapter 4, so little need be added here. Like GOMS, TAG is a method of describing interfaces, in terms of the linguistic structure of the commands. TAG is concerned specifically with interface consistency rather than user performance in general. It does, however, share many of the same problems that GOMS faces. Particularly, TAG requires an existing interface specification for evaluation and has little to say about non-expert performance. Its primary purpose is to reveal the internal inconsistency that might exist in the interface mechanisms themselves.

iii) CCT (Cognitive Complexity Theory)

The assumptions behind CCT (Kieras and Polson, 1985; Polson, 1987) are first, that interface skill can be described purely in terms of independent production rules (a form of rules which was described in Chapter 6), and second, that ease of learning is predictable from the number of rules to be learned. Hence, the less consistent an interface, the more rules are needed to describe it, and the harder it is to learn. Therefore, whilst GOMS and TAG model only expert performance, Kieras and Polson's (1985) CCT allows predictions of ease of learning to be generated for different devices based upon an analysis of each interface in terms of the number of production rules that are required to capture its functionality. The CCT approach does lead to clear predictions. For example, in the use of two applications such as a word processor and a drawing package, one can count the production rules required to model expertise. From this the transfer of skill in one interface to the other is estimated in terms of the number of productions required to gain the new skill and the number which can be carried over from experience with other interface.

As an example of the application of CCT, consider an experiment by Ziegler, Vossen and Hoppe (1990). They trained groups of users to carry out four tasks with graphics and text editors. They estimated that the skills to be learned to carry out these tasks could be described by 82 production rules. The four new tasks were presented to different groups of subjects in four sessions in such a way that each group was required to assimilate these rules in a different order. Ziegler *et al.* predicted, reasonably successfully, that the learning times in seconds for each of these tasks was a direct function of the number of rules to be learned in each session according to the formula:

Time to learn task = 199.5 + 17.2 * (number of new rules to be learned)

In CCT, the basic assumption is that expertise is quantifiable in terms of independent propositions. This sets an agenda for the experiment which concentrates upon interface complexity (measured by the number of propositions) and the transfer of training from task to another. This means that any one proposition should be equally easily learned regardless of whatever else the user knows, or the order in which they learn it. Scrutiny of Zeigler *et al.*'s experiment leaves one in doubt as to whether this criterion is actually met. Why, for example, does the equation above require 199.5 seconds to be added as a constant? In practice, users in two of Zeigler *et al.*'s experimental conditions took a considerable period of time to learn tasks which they ostensibly already knew how to carry out. The need to learn how to carry out these tasks in the context of a new task argues against the notion that propositions are abstract, independent and context-free units of knowledge. If the basic assumptions of CCT are questioned, it is not clear what the remaining value of the theory is.

iv) PUMs (Programmable User Models)

Some research into formal models is aimed at addressing some of the problems identified in the GOMS approach, namely the emphasis upon expert, error-free performance, and the requirement to specify methods of interface use. PUMs (Young, Green and Simon, 1989) is an attempt to generate a model of the user which can be 'instructed' to carry out tasks. The key element in this is that the PUM does not prescribe exactly how to carry out a given task. It uses in-built psychological principles and constraints, combined with some knowledge of the task and interface to be simulated, to solve the problem of how to perform the task.

The key element of research into PUMS is to identify these 'in-built psychological principles and constraints'. The perceived advantage of this approach, should it prove workable, is that it makes no assumptions about how users go about solving a task, and allows for the prediction of conceptual errors which other approaches cannot do. Hence, in principle, a designer can evaluate a new interface by letting a PUM loose upon it and examine the ensuing usability errors. In contrast to GOMS, TAG and CCT, which aim at quantifications of execution times and interface complexity, PUMs simulate users' behaviour, and in so doing alert the designer to unforeseen responses to interface designs.

To date, PUMs have shown some limited success (e.g. Young and Whittington, 1990; Howes and Young, 1991). It is fair to say that they are still at an experimental stage. Whether they can be developed to the point of being of wide value as an evaluation technique by others remains to be seen.

Walkthrough approaches

In response to the difficulty of applying formal and theoretical approaches to evaluation, 'walkthrough' approaches have been proposed. Walkthrough approaches are structured methods, based upon theory, whose purpose is to allow evaluators to evaluate interfaces by 'stepping' through their use without the need for formal modelling or usability testing. To exemplify this approach, we consider the 'cognitive walkthrough' approach of Lewis, Polson, Wharton and Rieman (1990).

The theoretical background to Lewis *et al.*'s approach is a model of exploratory learning, CE+, based upon CCT, and similar to ACT*. They emphasize the importance of 'guessing' as part of interface use. By guessing, they mean the ability to infer appropriate actions at interfaces from the available (usually visual) information; a process identified by Mayes *et al.* (1988) and which others have called display-based problem solving (e.g. Larkin, 1989). This was discussed in Chapter 8 of this book in relation to skill-as-exploration. This leads them to basic design principles such as (from Lewis *et al.* 1990):

1. Make the repertory of available actions salient.
2. Provide an obvious way to undo actions.
3. Offer few alternatives.
4. Require as few choices as possible.

The walkthrough itself comprises forms of structured questions, as exemplified in Table 10.6. Even from so short an extract, the bureaucracy of this approach is clear. It will also be obvious that the construction of these questions and their understanding is not actually simple. For example, what is meant by an identifier location?

It is questionable whether cognitive walkthroughs actually achieve the aim of providing a more accessible theoretical input to the evaluation process. The approach is too new to draw any firm conclusions, but hints in the literature suggest not. For example, on the basis of avoiding the difficult and time-consuming aspects of walkthroughs, Rowley and Rhoades (1992) advocate a less formal process of cognitive jogthroughs. Remarks from Wharton, Bradford, Jeffries and Franzke (1992) are also revealing. For example: "We found the looser application of the method ... to be more successful" (p. 385) ; and that the method was "not [readily] applicable without substantial extensions" (p. 387). These remarks suggest that the role of discretion in designing and

Table 10.6 An excerpt from a Walkthrough Form from Wharton *et al.* (1992). This short example illustrates the problem that ensuring procedural thoroughness through documentation is inevitably bureaucratic.

. . .

Step [B] Choosing the Next Correct Action:

[B.1] Correct Action: Describe the action that the user should take at this step in the sequence

[B.2] Knowledge Checkpoint: If you have assumed user knowledge or experience, update the USER ASSUMPTION FORM.

[B.3] System State Checkpoint: If the system state may influence the user, update the SYSTEM STATE FORM.

[B.4] Action Availability: Is it obvious to the user that this action is a possible choice here? If not, indicate why.

How many users might miss this action (% 100 75 50 25 10 5 0)?

[B.5] Action Identifiability:

[B.5.a] Identifier Location, Type, Wording, and Meaning:

_____No identifier is provided. (skip to subpart [B.5.d].)

Identifier Type: Label Prompt Description Other(Explain)

Identifier Wording_____

Is the identifier's location obvious? If not, indicate why.

[B.5.b] Link Between Identifier and Action: Is the identifier clearly linked with this action? If not, indicate why.

How many users won't make this connection (% 100 75 50 25 10 5 0)?

[B.5.c] Link between Identifier and Goal: Is the identifier easily linked with an active goal? If not indicate why.

How many users won't make this connection (% 100 75 50 25 10 5 0)?

. . .

carrying out walkthroughs is important. This undermines the objective credibility of the approach.

Heuristic evaluation

Heuristic evaluation consists of the application, by evaluators of varying degrees of expertise, of a set of heuristics to judge the adequacy of a design prototype.

The nine heuristics proposed by Molich and Nielsen (1990) are illustrated in Table 10.7. Their purpose is to be a stripped-down rule-base which is simple to apply. On the face of it, heuristic evaluation is hardly methodical:

> "Heuristic Evaluation is an informal method of usability analysis where a number of evaluators are presented with an interface and asked to comment upon it" (Nielsen and Molich, 1990).

Such evaluation is clearly unreliable. Indeed Nielsen and Molich (1990) estimate that individual evaluators found only between 20% and 51% of the problems to be identified, and that the correlations between evaluators was poor. Arguably, the whole purpose of the methods described above was to avoid such imprecision. Why then is heuristic evaluation advocated at all? A number of points lead to the conclusion that heuristic evaluation is a viable contender as methods go, providing the shortcomings are identified:

What is the real skill in evaluation?

A recurrent theme in other approaches to evaluation is their difficulty of practical use. Usability testing can require sophisticated experimental and statistical skills, and recent work also testifies to the same problems in GOMS, TAG, CCT and walkthrough approaches: Those people best at carrying out such approaches are already steeped in them; and even then they find it difficult. The same observation has already been made in relation to Task Analysis discussed in Chapter 9.

Table 10.7   These heuristics represent the distillation (and in some cases simplification) of the main issues in interface design. Their purpose is to provide as easily-remembered set of principles as possible to be used in practical situations. (From Nielsen and Molich, 1990).

Simple and natural language

Speak the user's language

Minimize user memory load

Be consistent

Provide feedback

Provide clearly market exits

Provide shortcuts

Good error messages

Prevent errors

Nielsen (1992) has emphasized the importance of expertise in finding usability problems by heuristic evaluation. One of the most important skills in evaluation is understanding the range of issues involved. All that is additionally required is sufficient method to ensure that the evaluation is structured and thorough. Arguably, the nine heuristics of heuristic evaluation provide this as well or better than other methods;

The added information from aggregating evaluations

One solution to the problem of variability between evaluators is to pool evaluators comments. Figure 10.2 shows the proportion of usability problems identified as a function of the expertise and number of evaluators used. For most identifiable problems, the chance of detecting problems increases

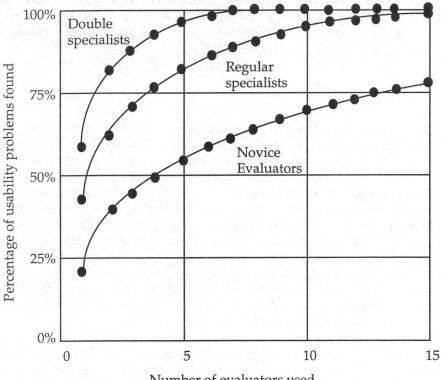

Figure 10.2 The average proportion of usability problems identified in heuristic evaluation as a function of the number of evaluators used and their interface expertise. 'Double' specialists are defined as interface experts with specific experience of the interfaces being evaluated (From Nielsen, 1992).

quickly as a function of the number of evaluators used and their expertise. Further, Nielsen and Landauer (1993) show that from such distributions it is possible to estimate the number of problems that remain to be identified. Such information is vital to designers because it provides an indication of how much more effort should be invested in the evaluation process.

Efficiency

A number of studies have recently been published in which different evaluation methods are compared. For example, Jeffries, Miller, Wharton and Uyeda (1991) compare usability testing, heuristic evaluation, guidelines, and cognitive walkthrough techniques in the evaluation of the same pre-release interface. Briefly, heuristic evaluation identified both the largest total number of problems and also the largest number of serious problems. The usability testing technique was more focused upon severe problems and less likely to throw up trivial problems, even if the number of serious problems it identified was smaller. The use of both guidelines and walkthroughs fared badly.

Cost-benefit trade-offs

Regardless of the efficiency of an evaluation technique, there is also the question of cost-benefit trade-offs. The relative efficiency of usability testing and heuristic evaluation suggests that different projects require different resources for adequate evaluation. A small project may get by with a few user trials or a small number of heuristic evaluators, but a large project may require extensive trialling or large numbers of skilled heuristic evaluators. On large projects, where large numbers of long-term users are anticipated, the costs of small design flaws are amplified. Both Nielsen and Landauer (1993) and Jeffries *et al.* (1991) have estimated the relationship between the cost of finding faults and the savings as a result of this cost for different evaluation methods. In both cases the outcome is the same: heuristic evaluation is more cost-effective than usability testing. For illustration, Nielsen and Landauer's estimates are shown in Table 10.8.

It could be argued that heuristic evaluation is simply the application of guidelines. As an evaluation approach, however, it very clearly out-performs guidelines. The heuristics used by Nielsen and others may well be more intrinsically effective than other guidelines, which are more detailed and numerous. However, the real strength of heuristic evaluation seems to lie in its recognition of the subjectivity of, and the variability between, evaluations. What Nielsen has provided in heuristic evaluation is a cost-effective method for minimizing the effects of these problems.

Table 10.8 Cost-benefit analyses for both user testing and heuristic evaluation in small, medium-large, and very large projects. In both cases the optimum number of test users or evaluators are used. The results show primarily that both evaluation techniques are highly cost-effective, as the right-hand columns show. This cost-effectiveness increases with the size of project concerned. Additionally, heuristic evaluation delivers greater cost-effectiveness than user testing. (From Nielsen and Landauer, 1993).

Project Size	Heuristic evaluation			Usability testing		
	Cost	Benefits	Benefit/ Cost ratio	Cost	Benefits	Benefit/ Cost ratio
Small	$9,400	$39,500	4.2	$10,000	$37,900	3.8
Medium-large	$13,600	$613,000	45	$18,000	$613,000	34
Very Large	$33,200	$8,200,000	247	$46,000	$8,200,000	178

10.4 Conclusions

This Chapter has concentrated upon the different reasons for which evaluation is carried out and the range of measures and methodologies which can be applied. Given the imprecise nature of the definitions of usability and evaluation, it is not surprising that philosophies of *evaluating usability* vary widely. The first conclusion to be drawn from this study is that evaluation is a subjective skill not a formal process.

Despite substantial theoretical and methodological scaffolding, techniques such as usability testing, GOMS analyses, and walkthroughs all serve a single common function: to orient the evaluator towards the interface in such a way as to systematize their evaluation. Beyond this function, in the worst case, the technical and intellectual complexities of certain approaches may only add to the cost of the evaluation.

We conclude this Chapter with what we believe to be two maxims about evaluation:

1) **Any approach which orients one to think about relevant issues of usability is likely to lead to useful insights**.

Therefore, almost all evaluation methods will improve interfaces to some extent. Protagonists for a particular evaluation technique cannot simply use interface 'improvement' as evidence for the validity of their approach because it applies to all evaluation methods.

2) **Evaluation methods cannot be compared or justified on theoretical grounds, but they can be assessed for their cost-effectiveness**

We suggest that theoretically-based evaluation methods provide only a weak coupling between theory and practice. An example of this is to be seen in Lewis *et al.*'s design principles for cognitive walkthroughs illustrated earlier. These design principles may well be consistent with CE +, but they are consistent with any number of other theories also. Put another way, these principles did not follow from CE +, and CE + was not needed to generate them. In the light of the first maxim above, merely showing that an evaluation method improves an interface is not evidence that the theoretical basis for it is justified. On the other hand, in the various comparisons of different methods, it has been demonstrated that some are much more cost-effective than others, that they differ in their needs for time and expertise to carry out, and in their ability to target attention upon serious problems.

These maxims are deliberately provocative; they are designed to make the point that evaluation is a practical matter in which economic issues are at least as important as theoretical ones. In practice, all the evaluation techniques we have described have advantages which will make them useful in certain circumstances. It is a matter of skill to determine which evaluation approaches to take in any particular case.

References

Adelson, B. (1981). Problem solving and the development of abstract categories in programming languages. *Memory and Cognition,* **9**, 422–433.

Adelson, B. (1992). Evocative agents and multi-media interface design. In P. Bauersfeld, J. Bennett, & G. Lynch (Eds.), *Proceedings of ACM Conference on Human Factors in Computing Systems,* (pp. 351–356). Monterey, California: ACM Press.

Alba, J. W., & Hasher, L. (1983). Is memory schematic? *Psychological Bulletin,* **93**, 203–231.

Allport, D. A., Antonis, B., & Reynolds, P. (1972). On the division of attention: A disproof of the single channel hypothesis. *Quarterly Journal of Experimental Psychology,* **24**, 225–235.

Allwood, C. M., & Eliasson, M. (1987). Analogy and other sources of difficulty in novices' very first text-editing. *International Journal of Man-machine Studies,* **27**, 1–22.

Anderson, J. R. (1983). *The architecture of cognition.* Cambridge, MA: Harvard University Press.

Anderson, J. R. (1987). Skill acquisition: Compilation of weak-method problem solutions. *Psychological Review,* **94**, 192–210.

Anderson, J. R. (1990). *Cognitive psychology and its implications.* New York: W. H. Freeman and Co.

Anderson, J. R., Farrell, R., & Sauers, R. (1984). Learning to program in Lisp. *Cognitive Science,* **8**, 87–129.

Annett, J., Duncan, K. D., Stammers, R. B., & Gray, M. J. (1971). *Task Analysis.* London: HMSO.

Anzai, Y., & Simon, H. A. (1979). The theory of learning by doing. *Psychological Review,* **86**, 124–140.

Apple (1987). *Human interface guidelines: The Apple desktop interface.* Reading MA: Addison-Wesley.

Arbib, M. A. (1977). *Computers and the cybernetic society.* N.Y.: Academic Press.

Ashworth, C., & Goodland, M. (1990). *SSADM: A practical approach.* London: McGraw-Hill.

Ausubel, D. P. (1968). *Educational psychology: A cognitive view.* New York: Holt, Rinehart and Winston.

Baddeley, A., & Hitch, G. (1974). Working memory. In G. H. Bower (Ed.), *The psychology of learning and motivation.* New York: Academic Press.

Baecker, R., & Small, I. (1990). Animation at the interface. In B. Laurel (Ed.), *The Art of Human-Computer Interface Design* (pp. 251–267). Reading, MA: Addison-Wesley.

Bainbridge, L. (1987). Ironies of automation. In J. Rasmussen, K. Duncan, & J. Leplat (Eds.), *New Technology and Human Error* (pp. 271–283). Chichester: Wiley.

Barnard, P. J., & Grudin, J. (1988). Command Names. In M. Helander (Ed.), *Handbook of human-computer interaction,* Elsevier Science Publishers B.V. (North-Holland).

Barnard, P. J., & Hammond, N. V. (1982). Usability and its multiple determination for the occasional user of interactive systems. In M. B. Williams (Ed.), *Pathways to the Information Society*, pp. 543–548. North-Holland.

Barnard, P. J., Hammond, N. V., Maclean, A., & Morton, J. (1982). Learning and remembering interactive commands in a text-editing task. *Behaviour and Information Technology*, **1**, 347–358.

Barnard, P. J., Hammond, N. V., Morton, J., & Long, J. B. (1981). Consistency and compatibility in human-computer dialogue. *International Journal of Man-Machine Studies*, **15**, 87–134.

Bartlett, F. C. (1932). *Remembering*. Cambridge: Cambridge University Press.

Beattie, G. W., Cutler, A., & Pearson, M. (1982). Why is Mrs Thatcher interrupted so often? *Nature*, **300** (5894), 744–747.

Benbasat, I., Dexter, A. S., & Todd, P. (1986). The influence of colour and graphical information presentation in a managerial decision simulation. *Human-Computer Interaction*, **2**, 65–92.

Benyon, D. (1992). The role of task analysis in systems design. *Interacting with Computers*, **4**, 102–139.

Berry, D. C., & Broadbent, D. E. (1984). On the relationship between task performance and associated verbalisable knowledge. *Quarterly Journal of Experimental Psychology*, **36A**, 209–231.

Biermann, A. W., Ballard, B. W., & Sigmon, A. H. (1983). An experimental study of natural language programming. *International Journal of Man-Machine Studies*, **18**, 71–87.

Black, J., & Moran, T. (1982). Learning and remembering command names. *Proceedings of the ACM Conference on Human Factors in Computing Systems*, (pp. 8–11). Gaithersburg, Maryland: ACM Press.

Black, J. B., Carroll, J. M., & McGuigan, S. M. (1987). What kind of minimal instruction manual is most effective. In J. M. Carroll & P. P. Tanner (Eds.), *Proceedings of the ACM Conference on Human Factors in Computing Systems*. Toronto: ACM Press.

Bowers, J. M., & Benford, S. D. (Eds.). (1992). *Studies in computer-supported cooperative work*. Amsterdam: Elsevier.

Bradford, J. H., Murray, W. D., & Carey, T. T. (1990). What kind of errors do Unix users make? In D. Diaper, D. Gilmore, G. Cockton, & B. Shackel (Eds.), *IFIP TC 13 Third International Conference on Human-Computer Interaction*, Cambridge, UK: North-Holland.

Bransford, J. D., & Johnson, M. K. (1972). Contextual prerequisites for understanding: Some investigations of comprehension and recall. *Journal of Verbal Learning and Verbal Behaviour*, **11**, 717–726.

Broadbent, D. E. (1963). *Perception and communication*. Oxford: Pergamon.

Brooke, J., Bevan, N., Brigham, F., Harker, S., & Youmans, D. (1990). Usability statements and standardisation - work in progress in ISO. In D. Diaper, D. J. Gilmore, G. Cockton & B. Shackel (Eds.), *Human-Computer Interaction: INTERACT '90*. Amsterdam: Elsevier.

Brooks, F. (1975). *The Mythical Man-Month*. Addison-Wesley.

Brown, I. D., Wastell, D. G., & Copeman, A. K. (1982). A psychophysiological investigation of system efficiency in public telephone switchrooms. *Ergonomics*, **25**, 1013–1040.

Brown, M. L., Newsome, S. L., & Glinert, E. P. (1989). An experiment into the use of auditory cues to reduce visual workload. *Proceedings of the ACM Conference on Human Factors in Computing Systems*, (pp. 339–346). Austin, Texas: ACM Press.

Bush, G., & Williams, E. (1978). Viewdata: The systematic development and testing of PO routing trees. UK Post Office Memorandum.

Buxton, W., Gaver, W., & Bly, S. (1989). The use of non-speech audio at the interface.

Proceedings of the ACM Conference on Human Factors in Computing Systems, Tutorial 10, Austin, Texas: ACM Press.

Byrne, M. D. (1993). Using icons to find documents. *Proceedings of the ACM Conference on Human Factors in Computing Systems,* (pp. 446–453). Austin, TX: ACM Press.

Cahill, M. C., & Carter, R. C. (1976). Colour code size for searching displays of different density. *Human Factors, 18,* 273–280.

Cakir, A., Hart, D. J., & Stewart, T. F. M. (1980). *Visual Display Terminals.* Chichester: Wiley.

Capindale, R. A., & Crawford, R. G. (1990). Using a natural language interface with casual users. *International Journal of Man-Machine Studies, 32,* 341–361.

Caplan, L. J., & Schooler, C. (1990). Problem solving by reference to rules or previous episodes: the effects of organized training, analogical models, and subsequent complexity of experience. *Memory and Cognition, 18,* 215–227.

Card, S. K. (1982). User perceptual mechanisms in the search of computer command menus. *Proceedings of the ACM Conference on Human Factors in Computing Systems* (pp. 190–196). Gaithersburg, MD: ACM Press.

Card, S. K., English, W. K., & Burr, B. J. (1978). Evaluation of mouse, rate-controlled isometric joystick, step keys, and text keys for text selection on a CRT. *Ergonomics, 21,* 601–613.

Card, S. K., Moran, J. P., & Newell, A. (1983). *The psychology of human computer interaction.* Hillsdale, NJ: Erlbaum.

Carey, M. S., Stammers, R. B., & Astley, J. A. (1989). Human-computer interaction design: The potential and pitfalls of Hierarchical Task Analysis. In D. Diaper (Ed.), *Task analysis for human-computer interaction.* (pp. 56–74). Chichester: Ellis Horwood.

Carroll, J. M. (1982a). Learning, using and designing filenames and command paradigms. *Behaviour and Information Technology, 1,* 327–346.

Carroll, J. M. (1982b). The adventure of getting to know a computer. *Computer, 15,* 49–61.

Carroll, J. M. (1990). *The Nurnberg funnel: Designing minimalist instruction for practical computer skill.* Cambridge, MA: MIT Press.

Carroll, J. M., & Campbell, R. L. (1989). Artifacts as psychological theories: The case of human-computer interaction. *Behaviour and Information Technology, 8,* 247–256.

Carroll, J. M., & Kellogg, W. A. (1989). Artifact as theory-nexus: Hermeneutics meets theory-based design. *Proceedings of the ACM Conference on Human Factors in Computing Systems,* (pp. 7–14). New York: ACM Press.

Carroll, J. M., Kellogg, W. A., & Rosson, M. B. (1990). The task-artifact cycle (No. RC 15731). IBM Research Division.

Carroll, J. M., & Mack, R. L. (1985). Metaphor, computing systems and active learning. *International Journal of Man-Machine Studies, 22,* 39–57.

Carroll, J. M., Smith-Kerker, P. L., Ford, J. R., & Mazur-Rimetz, S. A. (1987). The mininimal manual. *Human-Computer Interaction, 3,* 123–153.

Chapanis, A. (1991). Evaluating usability. In B. Shackel & S. J. Richardson (Eds.), *Human factors for infomatics usability* (pp. 359–396). Cambridge: Cambridge University Press.

Chapanis, A., Ochsman, R., Parrish, R. & Weeks, G. (1972). Studies in interactive communication: the effects of four communication modes on the behaviour of teams during cooperative problem solving. *Human Factors, 14,* 487–509.

Chase, W. G., & Simon, H. A. (1973). Perception in chess. *Cognitive Psychology, 4,* 55–81.

Cheng, P. W., Holyoak, K. J., Nisbett, R. E., & Oliver, L. M. (1986). Pragmatic versus syntactic appraoches to training deductive reasoning. *Cognitive Psychology, 18,* 293–328.

Chomsky, N. (1965). *Aspects of the theory of syntax.* Cambridge, MA: MIT Press.

Clegg, C., Warr, P., Green, T. R. G., Monk, A., Kemp, N., Allison, G., & Lansdale, M. W. (1988). *People and computers: How to evaluate your company's new technology.* Chichester, UK: Ellis Horwood.

Cole, I., Lansdale, M. W., & Christie, B. (1985). Dialogue design guidelines. In B. Christie (Ed.), *Human factors of Information Technology in the Office* (pp. 212–241). Chichester, UK: Wiley.

Coleman, L. M., & DePaulo, B. M. (1991). Uncovering the human spirit: Moving beyond disability and 'missed' communication. In N. Coupland, H. Giles, & J. M. Wiemann (Eds.), *"Miscommunication" and Problematic Talk* (pp. 61–84). Newbury Park, CA: Sage.

Conway, M. A., & Kahney, H. (1987). Transfer of learning in inference problems: Learning to program recursive functions. In J. Hallam & C. Mellish (Eds.), *Advances in artificial intelligence.* Chichester: Wiley.

Crossman, E. R. F. W. (1974). *Automation and skill.* London: Taylor and Francis.

Curtis, B., Krasner, H., & Iscoe, N. (1988). A field study of the software design process for large systems. *Communications of the ACM,* **31,** 1268–1287.

Cushman, W. H., & Crist, B. (1987). Illumination. In G. Salvendy (Ed.), *Handbook of Human Factors* (pp. 670–695). New York: Wiley.

Damodaran, L., Simpson, A., & Wilson, P. A. (1980). *Designing systems for people.* Manchester, UK: NCC Publications.

Davies, S. P. (1990). The nature and development of programming plans. *International Journal of Man-Machine Studies,* **32,** 461–481.

Davies, S. P. (In Press). Expertise in computer programming: The role of working memory and display-based problem solving. *International Journal of Man-Machine Studies.*

DeGroot, A. D. (1965). *Thought and choice in chess.* New York: Basic Books.

Détienne, F. (1990). Expert programming knowledge: a schema-based approach. In J. M. Hoc, T. R. G. Green, R. Samurçay and D. J. Gilmore (Eds.) *Psychology of Programming.* London: Academic Press.

Diaper, D. (Ed.). (1989). *Task analysis for human-computer interaction.* Chichester: Ellis Horwood.

Diaper, D., & Johnson, P. (1989). Task analysis for knowledge descriptions: theory and application in training. In J. Long & A. Whitefield (Eds.), *Cognitive Ergonomics and Human-Computer Interaction* (pp. 191–224). Cambridge: Cambridge University Press.

Dillon, A., McKnight, C., & Richardson, S. J. (1988). Reading from paper versus reading from screens. *The Computer Journal,* **31,** 457–464.

Doane, S. M., Pellegrino, J. W., & Klatsky, R. L. (1990). Expertise in a computer operating system: conceptualization and performance. *Human-Computer Interaction,* **5,** 267–304.

Douglas, S. A., & Moran, T. P. (1983). Learning text editor semantics by analogy. *Proceedings of the ACM Conference on Human Factors in Computing Systems,* (pp. 207–211). New York: ACM Press.

Dowell, J., & Long, J. (1989). Towards a conception for an engineering discipline of human factors. *Ergonomics,* **32,** 1513–1535.

Draper, S. W. (1985). The nature of expertise in UNIX. In B. Shackel (Ed.), *Interact '84: First IFIP Conference on Human-Computer Interaction* (pp. 465–471). Amsterdam: North Holland.

Draper, S. W., & Jordan, P. W. (1993). Consistency and the prospects for predicting learning burdens. Article submitted for publication.

Draper, S. W., & Oatley, K. (1992). Action Centered Manuals or Minimalist Instruction? In P. Holt, A. N. Williams (Eds.), *Computers & Writing.* Oxford: Intellect Books.

Drummond, K., & Hopper, R. (1991). Misunderstanding and its remedies: Telephone miscommunication. In N. Coupland, H. Giles, & J. M. Wiemann (Eds.), *"Miscommunication" and Problematic Talk* (pp. 301–314). Newbury Park, CA, USA: Sage.

Du, Z., & McCalla, G. (1991). CBMIP – A Case-based mathematics instructional planner. In L. Birnbaum (Ed.), *The International Conference on the Learning Sciences*, (pp. 122–129). Illinois: Association for the Advancement of Computing in Education.

Eason, K. D. (1984). Towards the experimental study of usability. *Behaviour and Information Technology*, **3**, 133–143.

Eason, K. D. (1988). *Information technology and organisational change*. London: Taylor and Francis.

Easterbrook, S. (1991). Negotiation and the role of requirements specification. Report No. CSRP 197. School of Computing and Cognitive Science, University of Sussex.

Edwards, A. (1989). Soundtrack: an auditory interface for blind users. *Human-Computer Interaction*, **4**.

Eisenberg, M., Resnick, M., & Turbak, F. (1987). Understanding procedures as objects. In G. M. Olson, S. Sheppard, & E. Soloway (Eds.), *Empirical studies of programmers: Second workshop*. Washington, DC: Ablex.

Erhlich, K., & Soloway, E. (1984). An empirical investigation of tacit plan knowledge in programming. In J. C. Thomas & M. L. Schneider (Eds.), *Human factors in computer systems*. Norwood, NJ: Ablex.

Eysenck, M. W. (1977). *Human memory: Theory, research, and individual differences*. Oxford, UK: Pergamon Press.

Eysenck, M. W., & Keane, M. T. (1990). *Cognitive psychology: A student's handbook*. London: LEA.

Falzon, P. (1990). Human-computer interaction: Lessons from human-human communication. In P. Falzon (Ed.), *Cognitive Ergonomics* (pp. 51–68). London: Academic Press.

Findlay, J. M. (1981). Local and global influences on saccadic eye movements. In D. F. Fisher, R. A. Monty, & J. W. Senders (Eds.), *Eye Movements: Cognition and Visual Perception*. Hillsdale, NJ: Lawrence Erlbaum.

Fisher, D. L., & Tan, K. C. (1989). Visual displays: The highlighting paradox. *Human Factors*, **31**, 17–30.

Fisher, S. (1986). *Stress and strategy*. London: LEA.

Fisher, S. S. (1986). *Telepresence master glove controller for dexterous robotic end-effectors*. Proceedings of SPIE Cambridge Symposium on Optical and Optoelectronic Engineering, Cambridge, MA.

Fitch, R. N. (1984) Deriving Man-Computer Dialogue Design Guidelines: A Preliminary Analysis of Form-Filling. Unpublished M.Sc thesis, Loughborough, UK.

Fitter, M. (1979). Towards more "natural" interactive systems. *International Journal of Man-Machine Studies*, **11**, 339–350.

Fong, G. T., Krantz, D. H., & Nisbett, R. E. (1986). The effects of statistical training on thinking about everyday problems. *Cognitive Psychology*, **18**, 253–292.

Furnas, G. W., Landauer, T. K. Gomez, L. M., & Dumais, S. T. (1983). Statistical semantics: analysis of the potential performance of keyword systems. *Bell System Technical Journal*, **62**, 1753–1806.

Galer, M., Harker, S. D. P., & Ziegler, J. (Eds.). (1992). *Methods and Tools in User-Centred Design for Information Technology*. Amsterdam, The Netherlands: North-Holland.

Gallaway, G. R. (1981). *Response Times to User Activities in Interactive Man/Machine Computer Systems*. Report HFP 81-25. NCR Corporation.

Gaver, W. W. (1989). The SonicFinder: An interface that uses auditory icons. *Human-Computer Interaction*, **4**(1).

Gaver, W. W. (1991). Technology affordances *Proceedings of the ACM Conference on Human Factors in Computing Systems*, (pp. 79–84). New Orleans, LA: ACM Press.

Gentner, D., & Gentner, D. R. (1983). Flowing waters or teeming crowds: Mental models of electricity. In D. Gentner & A. L. Stevens (Eds.), *Mental models*. New Jersey: LEA.

Gentner, D., & Stevens, A. L. (Ed.). (1983). *Mental models*. Hillsdale, NJ: Lawrence Erlbaum.

Gibson, J. J. (1966). *The senses considered as perceptual systems*. Boston: Houghton Mifflin.

Gibson, J. J. (1986). *The ecological approach to visual perception*. Hillsdale, NJ: Lawrence Erlbaum.

Gick, M. L., & Holyoak, K. J. (1983). Schema induction and analogical transfer. *Cognitive Psychology, 15*, 1–38.

Gilmore, D. J., & Green, T. R. G. (1988). Programming plans and programming expertise. *Quarterly Journal of Experimental Psychology, 40*A, 423–442.

Gladden, G. R. (1982). Stop the life-cycle, I want to get off. *ACM Software Engineering Notes, 7*(2), 29–32.

Good, M. (1985). The use of logging data in the design of a new text editor. In *Proceedings of the ACM Conference on Human Factors in Computing Systems,* (pp. 93–97). San Francisco, California: ACM Press.

Grandjean, E. (1987). Design of VDT Workstations. In G. Salvendy (Ed.), *Handbook of Human Factors* (pp. 1359–1397). New York: Wiley.

Gray, W. D., John, B. E., & Atwood, M. E. (1992). The precis of project Ernestine, or, an overview of a validation of GOMS. In *Proceedings of ACM Conference on Human Factors in Computing Systems,* (pp. 307–312). Monterey, California: ACM Press.

Green, A. J. K., & Gilhooly, K. J. (1990). Individual differences and effective learning procedures: the case of statistical computing. *International Journal of Man-Machine Studies, 33*, 97–119.

Green, T. R., Payne, S. J., Gilmore, D. J., & Mepham, M. (1984). Predicting expert slips. In *Interact '84, 1* (pp. 92–97). London: Elsevier.

Green, T. R. G. (1977). Conditional program statements and their comprehensibility to professional programmers. *Journal of Occupational Psychology, 50*, 93–109.

Green, T. R. G. (1980). Programming as a cognitive activity. In H. T. Smith & T. R. G. Green (Eds.), *Human interaction with computers*. London: Academic Press.

Green, T. R. G. (1989). Cognitive dimensions of notations. In A. Sutcliffe & L. Macaulay (Eds.), *People and Computer V*. Cambridge: Cambridge University Press.

Greenberg, S. (Ed.). (1991). *Computer-supported Cooperative Work and Groupware*. London: Academic Press.

Gregory, R. L. (1973). The confounded eye. In R. L. Gregory & E. H. Gombrich (Eds.), *Illusion in nature and art*. Duckworth.

Grice, H. P. (1967). Logic and conversation. In P. Cole, J. L. Morgan (Eds.) *Studies in Syntax, Vol. III*. New York: Seminar Press.

Grudin, J. (1989). The case against user interface consistency. *Communications of the ACM, 32*(10), 1164–1173.

Hammond, N. V., & Allinson, L. (1989). Extending hypertext for learning: An investigation of access and guidance tools. In A. Sutcliffe & L. Macaulay (Eds.), *People and Computers V*. Cambridge: Cambridge University Press.

Hammond, N. V., Hinton, G., Barnard, P. J., Maclean, A., Long, J., & Whitefield, A. (1984). Evaluating the interface of a document processor: A comparison of expert judgement and user observation. In *Interact '84, 2* (pp. 135–139). London: Elsevier Science Publishers.

Harker, S. (1987). Rapid prototyping as a tool for user centered design. In G.

Salvendy (Ed.), Cognitive Engineering in the Design of Human-Computer Interaction and Expert Systems: *Proceedings of the Second International Conference on Human-Computer Interaction* (pp. 365–372). Amsterdam: Elsevier Science.

Helander, M. (Ed.). (1990). *Handbook of Human-Computer Interaction.* Amsterdam: North-Holland.

Helander, M., Moody, T. S. A., Joost, M. G. (1990). Systems design for automated speech recognition. In M. Helander (Ed.) *Handbook of Human–Computer Interaction.* Amsterdam: North Holland.

Hewitt, T. T. (1989). Towards a rapid propotyping environment for interface design: Desirable features suggested by the electronic spreadsheet. In A. Sutcliffe & L. Macaulay (Eds.), *People and Computers V* (pp. 305–314). Cambridge University Press.

Hoare, C. A. R. (1981). The Emperor's old clothes. *Communications of the ACM, 24,* 75–83.

Holland, J. H., Holyoak, K. J., Nisbett, R. E., & Thagard, P. R. (1986). *Induction: Processes of inference, learning and discovery.* Cambridge, MA: MIT Press.

Howes, A., & Payne, S. J. (1990). Display-based competence: Towards user models for menu-driven interfaces. *International Journal of Man-Machine Studies, 33,* 637–655.

Howes, A., & Young, R. M. (1991). Predicting the learnability of Task-Action mappings. In *Proceedings of the ACM Conference on Human Factors in Computing Systems,* (pp. 113–118). New Orleans, LA: ACM Press.

Hutchins, E. L., Hollan, J. D., & Norman, D. A. (1986). Direct manipulation interfaces. In D. A. Norman & S. W. Draper (Eds.), *User centered system design.* Hillsdale, NJ: Lawrence Earlbaum.

IBM (1989). *Common user access: Advanced interface design guide* (SC26–4582). Boca Raton FL: International Business Machines Corporation.

Jacob, R. J. K. (1990). What you look at is what you get: Eye movement-based interaction techniques. In *Proceedings of the ACM Conference on Human Factors in Computing Systems,* (pp. 11–18). Seattle, Washington: ACM Press.

James, W. (1890). *Principles of psychology; Vol 1.* New York: Holt.

Jarke, M., Turner, J. A., Stohr, E. A., Vassiliou, Y., White, N. H., Michielsen, K. (1985). A field evaluation of natural language for data retrieval. *IEEE Transactions on Software Engineering* SE–11, **1,** 97–113.

Jeffries, R., Miller, J. R., Wharton, C., & Uyeda, K. M. (1991). User interface evaluation in the real world: A comparison of four techniques. In *Proceedings of the ACM Conference on Human Factors in Computing Systems,* (pp. 119–124). New Orleans, LA: ACM Press.

Jellinek, H. D., & Card, S. K. (1990). Powermice and user performance. In *Proceedings of the ACM Conference on Human Factors in Computing Systems,* (pp. 213–220). Seattle, Washington: ACM Press.

Johnson, W. L., & Soloway, E. (1985). PROUST: Knowledge-based program understanding. *IEEE Transactions on Software Engineering SE11,* **3,** 267–275.

Johnson-Laird, P. N. (1983). *Mental models.* London: Cambridge University Press.

Jorgensen, A. H., Barnard, P. J., Hammond, N. V., & Clark, I. A. (1983). Naming commands: An analysis of designers' naming behaviour. In T. R. G. Green, S. J. Payne, & G. C. van des Veer (Eds.), *The Psychology of Computer Use.* London: Academic Press.

Kamouri, A. L., Kamouri, J., & Smith, K. H. (1986). Training by exploration: Facilitating the transfer of procedural knowledge through analogical reasoning. *International Journal of Man-Machine Studies,* **24,** 171–192.

Kay, D. S., & Black, J. B. (1985). The evolution of knowledge representations with increasing expertise in using systems. In J. M. Carroll (Ed.), *Interfacing Thought.* Cambridge, MA: MIT Press.

Kelley, J. F. (1983). An empirical methodology for writing user-friendly natural-language computer applications. In *Proceedings of the ACM Conference on Human Factors in Computing Systems*, (pp. 193–196). Boston, MA: ACM, New York.

Kennedy, A., & Murray, W. S. (1991). The effects of flicker on eye movement control. *Quarterly Journal of Experimental Psychology*, **43**A(1), 79–100.

Kieras, D. E. (1988). Towards a paractical GOMS model methodology for user interface design. In M. Helander (Ed.), *Handbook of human-computer interaction*. North-Holland: Elsevier.

Kieras, D. E., & Bovair, S. (1984). The role of a mental model in learning to operate a device. *Cognitive Science*, **8**, 255–273.

Kieras, D. E., & Polson, P. G. (1985). An approach to the formal analysis of user complexity. *International Journal of Man-Machine Studies*, **22** (365–394).

Kim, J., & Lerch, F. J. (1992). Towards a model of cognitive processes in logical design: Comparing object-oriented and traditional functional decomposition software methodologies. In *Proceedings of the ACM Conference on Human Factors in Computing Systems*, Monterey, CA: ACM Press.

Kirwan, B., & Ainsworth, L. K. (Eds.). (1992). *A guide to task analysis*. London: Taylor and Francis.

Laming, D. R. J. (1986). *Sensory Analysis*. London: Academic Press.

Landauer, T. K., Calotti, K. M., & Hartwell, S. (1983). Natural command names and initial learning. *Communications of the ACM*, **26**, 495–503.

Lansdale, M. W. (1985). Beyond dialogue guidelines: The role of mental models. In B. Christie (Ed.), *Human Factors of Information Technology in the Office* (pp. 242–272). Chichester: Wiley.

Lansdale, M. W. (1988). On the memorability of icons in an information retrieval task. *Behaviour and Information Technology*, **7**(2), 131–151.

Lansdale, M. W., & Edmonds, E. (1992). Using memory for events in the design of personal filing systems. *International Journal of Man-Machine Studies*, **36**, 97–126.

Lansdale, M. W., Jones, M. R., & Jones, M. A. (1989). Visual search in iconic and verbal interfaces. In E. D. Megaw (Ed.), *Contemporary Ergonomics*. London: Taylor & Francis.

Lansdale, M. W., & Newman, I. (1991). System specification and usability. In K. Legge, C. Clegg, & N. Kemp (Eds.), *Case studies in information technology, people and organisations*. Oxford: NCC Blackwell.

Larkin, J. H. (1989). Display-based problem solving. In D. Klahr & K. Kotovsky (Eds.), *Complex Information Processing* (pp. 319–341). Hillsdale, NJ: Lawrence Erlbaum.

Ledgard, H., Whiteside, J., Singer, A., & Seymour, W. (1980). The natural language of interactive systems. *Communications of the ACM*, **23**, 556–563.

Lee, E., Whalen, T., McEwen, S., & Latremouille, S. (1984). Optimizing the design of menu pages for information retrieval. *Ergonomics*, **27**(10), 1051–1069.

Lewis, C., & Norman, D. A. (1986). Designing for error. In D. A. Norman & S. W. Draper (Eds.), *User Centred System Design*. Hillsdale, NJ: Lawrence Earlbaum.

Lewis, C., Polson, P. G., Wharton, C., & Rieman, J. (1990). Testing a walkthrough methodology for theory-based design of walk-up-and-use interfaces. In *Proceedings of the ACM Conference on Human Factors in Computing Systems* (pp. 235–242). Seattle, Washington: ACM Press.

Luff, P., Gilbert, N., & Frohlich, D. (Eds.). (1990). *Computers and Conversation*. London: Academic Press.

MacGregor, J. N., & Lee, E. S. (1987). Performance and preference in videotex menu retrieval: A review of the empirical literature. *Behaviour and Information Technology*, **6**(1), 43–68.

Mackay, W. E., Malone, T. W., Crowston, K., Rao, R., Rosenblitt, D., & Card, S. K. (1989).

How do experienced Information Lens users use rules? In *Proceedings of the ACM Conference on Human Factors in Computing Systems*, (pp. 211–216). Austin, TX: ACM Press.

Macleod, M., & Tillson, P. (1990). Pull-down, hold-down, or stay down? A theoretical and empirical comparison of three menu designs. In D. Diaper, D. J. Gilmore, G. Cockton, & B. Shackel (Eds.), *Interact '90*. Cambridge, UK: North-Holland.

Maguire, M. (1982). An evaluation of published recommentations on the design of man-computer dialogues. *International Journal of Man-Machine Studies, 16*, 237–261.

Mané, A. M., Adams, J. A., & Donchin, E. (1989). Adaptive and part-whole training in the acquisition of a complex perceptual-motor skill. *Acta Psychologica, 71*, 179–196.

Mani, K., & Johnson-Laird, P. N. (1982). The mental representation of spatial descriptions. *Memory and Cognition, 10*, 181–187.

Mannes, S. M., & Kinstch, W. (1991). Routine computing tasks: Planning as understanding. *Cognitive Science, 15*, 305–342.

Manktelow, K. I., & Jones, J. (1987). Principles from the psychology of thinking and mental models. In M. M. Gardiner & B. Christie (Eds.), *Applying Psychology to User-Interface Design*. Chichester: Wiley.

Matarazzo, J. D., Ulett, G. A., & Saslow, G. (1955). Human maze performance as a function of increasing levels of anxiety. *Journal of General Psychology, 53*, 79–95.

Mayer, R. E. (1975). Different problem-solving competencies established in learning computer programming with and without meaningful models. *Journal of Educational Psychology, 67*(6), 725–734.

Mayer, R. E. (1976). Comprehension as affected by structure of problem representation. *Memory and Cognition, 4*(3), 249–255.

Mayer, R. E. (1981). The psychology of how novices learn computer programming. *Computer Surveys, 13*(1), 121–141.

Mayes, J. T., Draper, S. W., McGregor, A. M., & Oatley, K. (1988). Information flow in a user interface: The effect of experience and context on the recall of MacWrite sessions. In D. M. Jones & R. Winder (Eds.), *People and computers IV* (pp. 275–289). Cambridge: Cambridge University Press.

McFarland, R. A. (1953). *Human factors in air transportation*. New York: McGraw-Hill.

McKeithen, K. B., Reitman, J. S., Rueter, H. H., & Hirtle, S. C. (1981). Knowledge organization and skill differences in computer programmers. *Canadian Journal of Psychology, 13*, 307–325.

McKendree, J., & Anderson, J. R. (1987). Effect of practice on knowledge and use of basic Lisp. In J. M. Carroll (Ed.), *Interfacing thought: Cognitive aspects of human-computer interaction*. Cambridge, MA: MIT Press.

Mezrich, J. J., Frysinger, S., & Slivjanovski, R. (1984). Dynamic representation of multivariate time series data. *Journal of the American Statistical Association, 79*, 34–40.

Michotte, A. (1963). *The perception of causality*, cited in S.K. Card, T.P. Moran and A. Newell *op. cit.*

Miller, G. A. (1956). The magical number seven plus or minus two: Some limits on our capacity for processing information. *Psychological Review, 63*, 81–97.

Minsky, M. (1977). Frame-section theory. In P. N. Johnson & P. C. Wason (Eds.), *Thinking: Readings in cognitive science*. Cambridge: Cambridge University Press.

Mohageg, M. F. (1991). Object-oriented versus bit-mapped graphics interfaces: Performance and preference differences for typical applications. *Behaviour and Information Technology, 10*, 121–148.

Molich, R., & Nielsen, J. (1990). Improving a human-computer dialogue. *Communications of the ACM, 33*(3), 338–348.

Mumford, E. (1979). The design of work: new approaches and new needs. In J. E. Rijnsdorp (Ed.), *Case Studies in Automation Related to the Humanisation of Work*. Oxford: Pergamon.

Myers, B. A. (1985). The importance of percent-done progress indicators for human-computer interfaces. In *Proceedings of the ACM Conference on Human Factors in Computing Systems,* (pp. 11–17). San Francisco, CA: ACM Press.

Myers, B. A., & Rosson, M. B. (1992). Survey on User Interface Programming. In P. Bauersfeld, J. Bennett, & G. Lynch (Eds.), *Proceedings of ACM Conference on Human Factors in Computing Systems.* Monterey, CA: ACM Press.

Mykowiecka, A. (1991). Text planning - how to make computers talk in natural-language. *International Journal of Man-Machine Studies,* **34,** 575–591.

Mynatt, B. T., Leventhal, L. M., Instone, K., Farhat, J., & Rohlman, D. (1992). Hypertext or Book: Which is better for answering questions? In *Proceedings of ACM Conference on Human Factors in Computing Systems,* Monterey, CA: ACM Press.

Neisser, U. (1976). *Cognition and reality.* San Francisco: W.H. Freeman.

Newell, A. (1973). *Production systems: Models of control structures.* New York: Academic Press.

Newell, A., & Simon, H. A. (1972). *Human problem solving.* Englewood Cliffs, NJ: Prentice-Hall.

Nielsen, J. (1990). *Hypertext and Hypermedia.* San Diego, CA: Academic Press.

Nielsen, J. (1992). Finding usability problems through heuristic evaluation. In *Proceedings of the ACM Conference on Human Factors in Computing Systems* (pp. 373–380). Monterey, CA: ACM Press.

Nielsen, J., Frehr, I., & Nymand, H. O. (1991). The learnability of Hypercard as an object-oriented programming system. *Behaviour and Information Technology,* **10,** 111–120.

Nielsen, J., & Landauer, T. K. (1993). A mathematical model of the finding of usability problems. In *Proceedings of the ACM Conference on Human Factors in Computing Systems* (pp. 206–213). Amsterdam: ACM Press.

Nielsen, J., & Molich, R. (1990). Heuristic evaluation of user interfaces. In *Proceedings of the ACM Conference on Human Factors in Computing Systems* (pp. 249–256). Seattle, Washington: ACM Press.

Norman, D. A. (1981a). The trouble with UNIX. *Datamation* (November issue).

Norman, D. A. (1981b). Categorisation of action slips. *Psychological Review,* **88,** 1–15.

Norman, D. A. (1983). Some observations on mental models. In D. Gentner & A. L. Stevens (Eds.), *Mental Models.* Hillsdale, NJ: Lawrence Erlbaum.

Norman, D. A. (1986). Cognitive Engineering. In D. A. Norman & S. W. Draper (Eds.), *User centered system design: New perspectives in human-computer interaction.* Hillsdale, NJ: Lawrence Erlbaum.

Norman, D. A. (1988). *The psychology of everyday things.* New York: Basic Books.

Norman, D. A., & Draper, S. W. (1986). *User-centered system design: New perspectives in human-computer interaction.* New York: Lawrence Erlbaum.

Norman, D. A., & Shallice, T. (1986). Attention to action: willed and automatic control of behaviour. In R. J. Davidson, G. E. Schwartz, & D. Shapiro (Eds.), *Consciousness and self regulation: Advances in research & theory* (pp. 1–18). New York: Plenum Press.

Noyes, J. (1983). The QWERTY keyboard: A review. *International Journal of Man-Machine Studies,* **18,** 265–281.

Olle, T. W., Hagelstein, J., MacDonald, I. G., Rolland, C., Sol, H. G., Assche, F. V., & Verrijn-Stuart, A. (1988). *Information systems methodologies.* Wokingham: Addison-Wesley.

Olson, J. R., & Nilson, E. (1988). Analysis of the cognition involved in spreadsheet software interaction. *Human Computer Interaction,* **3,** 309–349.

Olson, J. R., & Olson, G. M. (1990). The growth of cognitive modeling in human computer interaction since GOMS. *Human Computer Interaction,* **5,** 221–266.

Ormerod, T. C., & Ball, L. J. (1993). Does programming knowledge or design strategy determine shifts of focus in Prolog programming? In C. R. Cook, J. C. Scholtz and J. C. Spohrer (Eds.) *Empirical Studies of Programmers '5*, Palo Alto, CA: Ablex.

Ormerod, T. C., & Jones, G. V. (1991). Acquisition of recursive programming skills in Prolog. *Poster presented at International Conference on the Learning Sciences,* Evanston, IL.

Ormerod, T. C., Manktelow, K. I., & Jones, G. V. (1993). Reasoning with three conditionals: Biases and mental modes. *Quarterly Journal of Experimental Psychology,* **46A,** 653–677.

Ormerod, T. C., Manktelow, K. I., Robson, E. H., & Steward, A. P. (1986). Content and representation effects with reasoning tasks in PROLOG form. *Behaviour and Information Technology,* **5,** 157–168.

Ormerod, T. C., Manktelow, K. I., Steward, A. P., & Robson, E. H. (1990). The effects of content and representation on transfer of PROLOG reasoning skills. In K. Gilhooly, M. T. Keane, R. Logie, & G. Erdos (Eds.), *Lines of thinking: Reflections on the psychology of human thought.* Chichester: Wiley.

Paap, K. R., & Roske-Hofstrand, R. J. (1988). Design of menus. In M. Helander (Ed.), *Handbook of Human-Computer Interaction* (pp. 203–236). Amsterdam: North-Holland.

Pachella, R. (1974). The interpretation of reaction time in information processing research. In B. Kantowitz (Ed.), *Human Information Processing: Tutorials in performance and cognition.* Hillsdale, NJ: Lawrence Erlbaum.

Palmiter, S., & Elkerton, J. (1991). An evaluation of animated demonstrations for learning computer-based tasks. In *Proceedings of the ACM Conference on Human Factors in Computing Systems* (pp. 257–263). New Orleans, LA: ACM Press.

Parnas, D. L., & Clements, P. C. (1986). A rational design process: How and why to fake it. *IEEE Transactions on Software Engineering,* **12,** 251–257.

Patrick, J. (1992). *Training: Research and Practice.* London: Academic Press.

Patterson, R. (1990). Auditory warning sounds in the work environment. *Phil. Trans. R. Soc. Lond., B* **327,** 485–492.

Payne, S. J. (1988). Methods and mental models in theories of cognitive skill. In J. Self (Ed.), *Artificial Intelligence and human learning.* London: Chapman and Hall.

Payne, S. J. (1991). A descriptive study of mental models. *Behaviour and Information Technology,* **10,** 3–21.

Payne, S. J., & Green, T. R. G. (1986). Task-Action Grammars: A model of mental representation of task languages. *Human-Computer Interaction,* **2,** 93–133.

Pearson, G., & Weiser, M. (1988). Exploratory evaluation of a planar foot-operated cursor-positioning device. In *Proceedings of the ACM Conference on Human Factors in Computing Systems* (pp. 13–18). Washington, DC: ACM Press.

Peck, V. A., & John, B. E. (1992). Browser-SOAR: A computational model of a highly interactive task. In *Proceedings of the ACM Conference on Human Factors in Computing Systems* (pp. 165–172). Monterey, CA: ACM Press.

Pollack, I. (1952). The information of elementary auditory displays 1. *Journal of the Acoustical Society of America,* **14,** 745–749.

Polson, P. G. (1987). A quantitative theory of human-computer interaction. In J. M. Carroll (Ed.), *Interfacing thought: Cognitive aspects of human-computer interaction.* Cambridge MA: Bradford Books/MIT Press.

Postman, L., & Underwood, B. J. (1973). Critical issues in interference theory. *Memory and Cognition,* **1,** 19–40.

Potosnak, K. M. (1988). Keys and keyboards. In M. Helander (Ed.), *Handbook of human-computer interaction.* Amsterdam: Elsevier.

Potter, R. L., Weldon, L. J., & Shneiderman, B. (1988). Improving the accuracy of touch screens: an experimental evaluation of three strategies. In *Proceedings of the ACM Conference on Human Factors in Computing Systems,* (pp. 27–32). Washington, DC: ACM Press.

Ravden, S., & Johnson, G. (1989). *Evaluating usability of human-computer interfaces.* Chichester: Ellis Horwood.

Reason, J. T. (1990). *Human Error.* Cambridge: Cambridge University Press.

Reason, J. T., & Mycielska, K. (1982). *Absent-minded? The psychology of mental lapses and everyday errors.* New Jersey: Prentice-Hall.

Reed, S. K., Dempster, A., & Ettinger, M. (1985). Usefulness of analogous solutions for solving algebra word problems. *Journal of Experimental Psychology: Learning, Memory and Cognition,* **11,** 106–125.

Reisner, P. (1990). What is inconsistency? In D. Diaper, D. J. Gilmore, G. Cockton, & B. Shackel (Eds.), *Human Computer Interaction: INTERACT '90.* Amsterdam: Elsevier.

Resnick, P., & Virzi, R. A. (1992). Skip and scan: Cleaning up telephone interfaces. In *Proceedings of the ACM Conference on Human Factors in Computing Systems* (pp. 419–426) Monterey, CA: ACM Press.

Riesbeck, C. K., & Schank, R. C. (1989). *Inside case-based reasoning.* Hillsdale, NJ: Lawrence Erlbaum.

Rigg, A., & Sandringham, S. J. (1977). *Telewhich Group Discussion Reports No. 2 and 3.* Consumer Association.

Rist, R. S. (1986). Plans in programming: definition, demonstration and development. In E. Soloway & S. Iyengar (Eds.), *Empirical studies of programmers.* Norwood, NJ: Ablex.

Rist, R. S. (1991). Knowledge creation and retrieval in program design. A comparison of novice and intermediate programmers. *Human Computer Interaction,* **6,** 1–46.

Roberts, T. L., & Moran, T. P. (1983). The evaluation of text editors: methodology and empirical results. *Communications of the ACM,* **26,** 265–283.

Rogers, Y., Rutherford, A., & Bibby, P. A. (Eds.). (1992). *Models in the mind.* London: Academic Press.

Rosson, M. B., & Alpert, S. R. (1990). The cognitive consequences of object-oriented design. *Human Computer Interaction,* **5,** 345–379.

Rouse, W., & Rouse, S. (1983). Analysis and classification of human error. *IEEE Transactions, Systems, Man & Cybernetics,* **13**(4), 539–549.

Rowley, D. E., & Rhoades, D. G. (1992). The cognitive jogthrough: A fast-paced user interface evaluation procedure. In *Proceedings of the ACM Conference on Human Factors in Computing Systems,* (pp. 389–395). Monterey, CA: ACM Press.

Rumelhart, D. E., & McClelland, J. L. (Eds.). (1986). *Parallel distributed processing: Explorations in the microstructure of cognition, Vol 4.* Cambridge, MA: Bradford Books/MIT Press.

Rutter, D. K. (1987). *Communicating by Telephone.* Oxford: Pergamon Press.

Salvendy, G. (Ed.). (1987). *Handbook of Human Factors.* Chichester: Wiley.

Schneider, W., & Shiffrin, R. M. (1977). Controlled and Automatic Human Information Processing. I. Detection, Search and Attention. *Psychological Review,* **84**(1), 1–55.

Searle, J. (1984). *Minds, brains, and science: the 1984 Reith lectures.* London: BBC Publications.

Sebrechts, M. M., & Marsh, R. L. (1989). Components of computer skill acquisition: some reservations about mental models and discovery learning. In G. Salvendy & M. J. Smith (Eds.), *Designing and Using Human-Computer Interfaces and Knowledge-Based Systems.* Netherlands: Elsevier.

Seligman, M. E. P. (1975). *Helplessness: On depression development and death.* San Francisco: W. H. Freeman.

Shackel, B. (1991). Usability – Context, Framework, Definition, Design and Evaluation. In B. Shackel & S. J. Richardson (Eds.), *Human Factors for Informatics Usability* (pp. 21–38). Cambridge: Cambridge University Press.

Shaffer, L. H. (1975). Multiple attention in continuous verbal tasks. In P. M. A. Rabbitt & S. Dornic (Eds.), *Attention and Performance V*. London: Academic Press.

Sharples, M., & O'Malley, C. (1988). A framework for the design of a writer's assistant. In J. Self (Ed.), *Artificial Intelligence and Human Learning*. London: Chapman Hall.

Shepherd, A. (1985). Hierarchical task analysis and training decisions. *Programmed Learning and Educational Technology*, **22**, 162–176.

Shepherd, A. (1986). Issues in the training of process operators. *International Journal of Industrial Ergonomics*, **1**, 49–64.

Shepherd, A. (1989). Analysis and training in information technology tasks. In *Task Analysis for Human-Computer Interaction* (pp. 15–55). Chichester: Ellis Horwood.

Shepherd, A. (1993). An approach to information requirements specification for process control tasks. *Ergonomics*, **36**, 805–817.

Shepherd, A., & Ormerod, T. C. (1991). Information and skill in process control tasks. In *Proceeding of ICSE '91* (pp. 122–134). Aston University, UK.

Shiffrin, R., & Schneider, W. (1977). Controlled and automatic human information processing: II. Perceptual learning, automatic attending, and a general theory. *Psychological Review*, **84**, 127–190.

Shneiderman, B. (1980). *Software psychology: Human factors in computer and information systems*. Boston, MA: Little, Brown.

Shneiderman, B. (1987). *Designing the user interface*. Reading, MA: Addison-Wesley.

Shneiderman, B., & Mayer, R. (1979). Syntactic/semantic interactions in programmer behaviour: a model and experimental results. *International Journal of Computer and Information Sciences*, **7**, 219–239.

Singley, M. K., & Anderson, J. R. (1985). The transfer of text-editing skill. *International Journal of Man-Machine Studies*, **3**, 223–274.

Siochi, A. C., & Hix, D. (1991). A study of computer-supported user interface evaluation using maximal repeating pattern analysis. In *Proceedings of the ACM Conference on Human Factors in Computing Systems* (pp. 301–305). New Orleans, LA: ACM Press.

Smith, D. C., Irby, C., Kimball, R., Verplank, B., & Harslem, E. (1982), Designing the Star interface. *Byte*, 242–282.

Smith, S., Bergeron, R. D., & Grinstein, G. G. (1990). Sterophonic and surface sound generation for exploratory data analysis. In *Proceedings of the ACM Conference on Human Factors in Computing Systems* (pp. 125–132). Seattle, WA: ACM Press.

Smith, S. L., & Mosier, J. N. (1984). *Design guidelines for user-system interface software*. No. ESD-TR-84-190). USAF Electronic Systems Division.

Soloway, E., & Erlich, K. (1984). Empirical studies of programming knowledge. *IEEE Transactions on Software Engineering*, **SE-10**, 595–609.

Soloway, E., Erlich, K., & Bonar, J. (1982). Tapping into tacit programming knowledge. In *Proceedings of ACM Conference on Human Factors in Computing Systems*, Gaithersburg, MD: ACM Press.

Spence, R., & Apperley, M. (1982). Database navigation: an office environment for the professional. *Behaviour and Information Technology*, **1**(1), 43–54.

Stammers, R. B. (1982). Part and whole practice in training for procedural tasks. *Human Learning*, **1**, 185–207.

Starker, I., & Bolt, R. A. (1990). A gaze-responsive self-disclosing display. In *Proceedings of the ACM Conference on Human Factors in Computing Systems* (pp. 3–9). Seattle, WA: ACM Press.

Stephens, M. (1980). *Three Mile Island*. London: Junction Books.

Stephens, M. A., & Bates, P. E. (1990). Requirements engineering by prototyping: Experiences in development of estimating system. *Information and Software Technology*, **32**(4), 253–257.

Stewart, T. F. M. (1990). SIOIS – Standard interfaces of interface standards. In D. Diaper, D. J. Gilmore, G. Cockton, & B. Shackel (Eds.), *Human-Computer Interaction (Interact '90)* (pp. xxix–xxxiv). North Holland: Elsevier Science Publishers B.V.

Stroop, J. R. (1935). Studies of interference in serial verbal reactions. *Journal of Experimental Psychology*, **18**, 643–662.

Suchman, L. A. (1987). *Plans and situated actions.* Cambridge: Cambridge University Press.

Sutcliffe, A. (1988). *Human-computer interface design.* London: Macmillan.

Takeuchi, A., & Nagao, K. (1993). Communicative facial displays as a new conversational modality. In *Proceedings of the ACM Conference on Human Factors in Computing Systems* (pp. 187–193). Amsterdam: ACM Press.

Taylor, J., & du Boulay, B. (1986). Studying novice programmers: Why they may find learning Prolog hard. In J. Rutkowska (Ed.), *Issues for developmental psychology.* New York: Wiley.

Teal, S. L., & Rudnicky, A. I. (1992). A performance model of system delay and user strategy selection. In *Proceedings of the ACM Conference on Human Factors in Computing Systems* (pp. 295–305). Monterey, CA: ACM Press.

Tetzlaff, L., & Schwartz, D. R. (1991). The use of guidelines in interface design. In *Proceedings of the ACM Conference on Human Factors in Computing Systems* (pp. 329–333). New Orleans, LA: ACM Press.

Thovtrup, H., & Nielsen, J. (1991). Assessing the usability of a user interface standard. In *Proceedings of the ACM Conference on Human Factors in Computing Systems* (pp. 335–341). New Orleans, LA: ACM Press.

Transport, The Department of (1990). *Report on the accident to Boeing 737–400 G-OBME near Kegworth, Leicestershire on 8 January 1989.* No. Air Accident Report 4/90. Air Accidents Investigation Branch.

Travis, D. (1991). *Effective colour displays.* London: Academic Press.

VanDijk, T. A., & Kintsch, W. (1983). *Strategies of discourse comprehension.* New York: Academic Press.

Vidgen, G., & Hepworth, J. (1990). Yesterday's philosophy. *British Journal of Healthcare Computing*, **7**.

Wason, P. C. (1961). Response to affirmative and negative binary statements. *British Journal of Psychology*, **52**, 133–142.

Waters, R. C. (1985). The programmer's apprentice: A session with KBEmacs. *IEEE Transactions on Software Engineering*, 1296–1320.

Weizenbaum, J. (1976). *Computer power and human reason: From judgement to calculation.* San Fransisco: W. H. Freeman.

Wertheimer, M. (1958). *Productive thinking - 2nd ed.* New York: Harper & Row.

Wharton, C., Bradford, J., Jeffries, R., & Franzke, M. (1992). Applying cognitive walkthroughs to more complex user interfaces: Experiences, issues and recommendations. In *Proceedings of the ACM Conference on Human Factors in Computing Systems* (pp. 381–388). Monterey, CA: ACM Press.

Whiteside, J., Bennett, J., & Holtzblatt, K. (1988). Usability enginerring: Our experience and evolution. In M. Helander (Ed.), *Handbook of human-computer interaction.* Amsterdam: Elsevier.

Whiteside, J., Jones, S., Levy, P. S., & Wixon, D. (1985). User Performance with Command, Menu and Iconic Interfaces. In *Proceedings of the ACM Conference on Human Factors in Computing Systems*, San Francisco, CA: ACM Press.

Wiedenbeck, S., & Scholtz, J. (1989). Beacons: A knowledge structure in program comprehension. In G. Salvendy & M. J. Smith (Eds.), *Designing and using human-computer interfaces and knowledge-based systems.* Amsterdam: Elsevier.

Wilde, N., & Lewis, C. (1990). Spreadsheet-based interactive graphics: From prototype to tool. In *Proceedings of the ACM Conference on Human Factors in Computing Systems* (pp. 153–159). Seattle, Washington: ACM Press.

Wilkins, A. J. (1984). Visual discomfort and cathode ray tube displays. In *Interact '84*. London.

Williges, B. H., & Williges, R. C. (1984). Dialogue design considerations for interactive computer systems. *Human Factors Review*, 167–211.

Woolley, B. (1992). *Virtual worlds - A journey in hype and hyperreality*. Oxford: Blackwell.

Wright, P. (1984). Informed design for forms. In R. Easterby & H. Zwaga (Eds.), *Information Design*. Chichester: Wiley.

Wu, Q., & Anderson, J. R. (In Press). Problem solving transfer among programming languages. *Human-Computer Interaction*.

Yang, Y. (1992). Motivation, practice and guidelines for 'undoing'. *Interacting with Computers*, 4(1), 23–40.

Young, R. M. (1981). The machine inside the machine: users' models of pocket calculators. *International Journal of Man-Machine Studies*, 15, 51–85.

Young, R. M. (1983). Surrogates and mappings: Two kinds of conceptual models for interactive devices. In D. Gentner & A. L. Stevens (Eds.), *Mental models* (pp. 35–52). Hillsdale, NJ: Lawrence Erlbaum.

Young, R. M., Green, T. R. G., & Simon, T. (1989). Programmable user models for predictive evaluation of interface designs. In *Proceedings of the ACM Conference on Human Factors in Computing Systems*, (pp. 15–19). Austin, TX: ACM Press.

Young, R. M., & Hull, A. (1982). Cognitive aspects of the selection of viewdata options by casual users. In M. B. Williams (Ed.), *6th International Conference on Computer Communication*. London: North-Holland.

Young, R. M., & Whittington, J. (1990). Using a knowledge analysis to predict conceptual errors in text-editor usage. In *Proceedings of the ACM Conference on Human Factors in Computing Systems* (pp. 91–97). Seattle, WA: ACM Press.

Zhang, J. (1991). The interaction of internal and external representations in a problem solving task. In *13th Annual Conference of the Cognitive Science Society* (pp. 954–958). Chicago, IL: Lawrence Erlbaum.

Ziegler, J. E., Vosson, P. H., & Hoppe, H. U. (1990). Cognitive Complexity of Human-Computer Interfaces. In P. Falzon (Ed.), *Cognitive Ergonomics* (pp. 27–38). London: Academic Press.

Zoltan-Ford, E. (1991). How to get people to say and type what computers can understand. *International Journal of Man-Machine Studies*, 34, 527–547.

Author Index

Subject Index